T0220961

An Introduction to
Statistical Modelling

An Introduction to Statistical Modelling

Wojtek J. Krzanowski
Professor of Statistics
University of Exeter
UK

John Wiley & Sons, Ltd

First published in Great Britain in 1998
This impression reprinted in 2002 by
Arnold, a member of the Hodder Headline Group
338 Euston Road, London NW1 3BH

Co-published in the United States of America by
Oxford University Press Inc.,
198 Madison Avenue, New York, NY 10016

© 1998 Wojtek J. Krzanowski

John Wiley & Sons Ltd, The Atrium, Southern Gate, Chichester, West
Sussex, PO19 8SQ, United Kingdom

For details of our global editorial offices, for customer services and
for information about how to apply for permission to reuse the copyright
material in this book please see our website at www.wiley.com.

The right of the author to be identified as the author of this work has
been asserted in accordance with the Copyright, Design and Patents Act
1988.

All rights reserved. No part of this publication may be reproduced, stored in
a retrieval system, or transmitted, in any for or by any means, electronic,
mechanical, photocopying, recording or otherwise, except as permitted by
the UK Copyright, Designs and Patents Act 1988, without the prior permission
of the publisher.

Wiley also publishes its books in a variety of electronic formats. Some content
that appears in print may not be available in electronic books.

Designations used by companies to distinguish their products are often
claimed as trademarks. All brand names and product names used in this
book are trade names, service marks, trademarks or registered trademarks
of their respective owners. The publisher is not associated with any product
or vendor mentioned in this book. This publication is designed to provide
accurate and authoritative information in regard to the subject matter covered.
It is sold on the understanding that the publisher is not engaged in rendering
professional services. If professional advice or other expert assistance is required,
the services of a competent professional should be sought.

British Library Cataloguing in Publication Data
A catalogue record for this book is available from the British Library

Library of Congress Cataloging-in-Publication Data
A catalog record for this book is available from the Library of Congress

ISBN 978-0-470-71101-9

Typeset in 10/12 pt Times by Focal Image, Torquay

Contents

Series preface

Arnold Texts in Statistics is a series that is designed to provide an introductory account of key subject areas in statistics. Each book will focus on a particular area, and subjects to be covered include regression analysis, time series, statistical inference and multivariate analysis. Texts in this series will combine theoretical development with practical examples. Indeed, a distinguishing feature of the texts is that they will be copiously illustrated with examples drawn from a range of applications. These illustrations will take full account of the widespread availability of statistical packages and other software for data analysis. The theoretical content of the texts will be sufficient for an appreciation of the techniques being presented, but mathematical detail will only be included when necessary. The texts are designed to be accessible to undergraduate and postgraduate students of statistics. In addition, they will enable statisticians and quantitative scientists working in research institutes, industry, government organizations, market research agencies, financial institutions, and so on, to update their knowledge in those areas of statistics that are of direct relevance to them.

David Collett
Series Editor

Preface

Statistical treatment of data can be roughly categorized as either *description* or *analysis*. In the former case we are primarily interested in summarizing the available data, highlighting their main features and drawing out any important patterns or trends. There are many statistical techniques that can help us to achieve this objective, ranging from simple summary calculations and various graphical and tabular displays to specialized methodology such as principal component analysis for multivariate data. However, description is by its very nature limited: it is concerned merely with the data presented, and nothing more. Analysis, on the other hand, has a wider brief. It assumes *inter alia* that the available data form only a subset of all the data that *might* have been collected, and then attempts to use the information in the available data to make more general statements about either the larger set or about the mechanism that is operating to produce the data.

In order to make such statements, we need first to abstract the essence of the data-producing mechanism into a form that is amenable to mathematical and statistical treatment. Such a formulation will typically involve mathematical equations that express relationships between measured variables, and assumptions about the random processes that govern the outcomes of individual measurements. This is the statistical *model* of the system. Fitting the model to a given set of data will then provide a framework for extrapolating the results to a wider context or for predicting future outcomes, and can often also lead to an explanation of the system.

Statistical modelling therefore plays a vital part in all practical applications of statistics. However, it can arise in a variety of forms and at a variety of levels of complexity and sophistication. An undergraduate programme in which statistics is either the sole or a major component will usually have a set of modules concerned with modelling that range from the elementary to the advanced, complementing other sets of modules that focus on theoretical and computational aspects of statistics. Such a 'modelling' stream of modules will typically cover the basic ideas of modelling and inference in the first year, will take a comprehensive look at regression and analysis of variance in the second year, and will include an advanced module built around the generalized linear model in the final year.

Many books now exist within this general field, but most focus on just a subset of the above topics and there are relatively few texts that cover the full range at an undergraduate level. Moreover, many are written from a mathematical or theoretical rather than a practical perspective. My aim in writing this volume has therefore been to provide a single reference which has an applied slant and which will cater for all three years of such a programme. To keep the size manageable I have concentrated on the main issues, ignoring peripheral (albeit interesting) questions. Moreover, I have adopted a practical approach in which only the most essential mathematical justifications are given in detail, with more complicated or esoteric mathematics omitted entirely. This ensures that attention is firmly focussed on the statistical aspects of the techniques, without any unnecessary mathematical distractions.

As an important constituent of such an approach, all illustrations of the derived techniques use real data sets. Some of these have come from previously published sources which are acknowledged in the text. Others have come from lecture and examination material produced by various colleagues; in this respect I am grateful to Suzanne Evans, John Hinde and Ken Read for permission to use some of their own data sets, and to Ian Wilson, Director of the Statistical Services Centre at the University of Reading, for permission to use material from various short courses run by that Centre.

In conclusion I would like to thank Dave Collett for reading some of the chapters at draft stage, and an anonymous reader engaged by Arnold who cast a critical eye over the whole manuscript. Their comments have been very helpful, but any errors that have survived these readings are of course entirely my own responsibility. Any further comments or suggestions for improvement from users of the text will be very welcome.

Wojtek Krzanowski
University of Exeter
January 1998

1
Introduction

1.1 Models in data analysis

We are all familiar with the idea of a model in everyday life, and can readily visualize a range of models of varying complexity and sophistication. As children most of us have played with model cars, or model trains, or model houses, for example. Some may even have carried this predilection into adult life and spent many hours painstakingly building such things as model aeroplanes as a hobby. However, models are also used more seriously in various walks of professional life. For instance, architects and town planners often exhibit models of proposed new buildings or other major developments in order to test the public's reaction to them; medical schools employ models of the brain to teach students neurology, or models of the body to teach anatomy; and complex or costly electronic or technological equipment will generally involve the use of models in the developmental stage. As examples of the last type, model aircraft subjected to a wind tunnel will demonstrate their robustness or otherwise to severe external forces, model submarines will have their performance capabilities examined in a laboratory tank, and model rockets launched at a test site will highlight any possible manufacturing defects.

So what can we say in general about such models? Clearly, they are all intended to be small-scale representations of specific physical entities. An important feature, however, is that they will all vary as regards the accuracy of these representations. The key point here is that a model needs only to be as accurate as necessary *for the task in hand*. The toy car is used simply as a trigger for the child's imagination and is therefore purely *representational*: only a fairly crude approximation to reality is sufficient, and much of the associated detail may either be absent or even incorrect without detracting from its effectiveness. The architect's or town planner's models are primarily *descriptive*: they are intended to convey the main features of the proposed developments so as to assess their possible impact on surrounding environments. Similarly, the medical student is interested in learning about the bodily system, so needs enough detail in order to gain adequate comprehension. These models aim to impart *an understanding of the proposed system*, so that more detail is necessarily called for than in a purely representational model. Nevertheless, perfect working accuracy is not essential, unlike the cases of the model aircraft, rocket or submarine. These models are being used to *predict* the behaviour of their real counterparts under adverse or testing conditions. Considerable detail is therefore needed for the simulation to be an accurate one – if it isn't sufficiently accurate then not only the tests but also a considerable amount of money will have been wasted, and the danger of adopting a faulty machine will have been incurred.

These are all physical models, but similar categorization carries over to *mathematical* models, albeit at a greater degree of abstraction. In a mathematical model, we focus on a particular aspect of a given system and attempt to find an equation or set of equations that describes this aspect. For example, we may observe the trajectory of a body in a particular medium (e.g. a planet in space, a stone dropped into a tank of still water in a laboratory, or a pebble skimmed across the surface of the sea from a beach), and attempt to find equations describing its motion. The accuracy of representation will once again be governed by the purposes to which the modelling is to be put, i.e. *description* versus *prediction*, but will additionally be influenced by the degree to which external factors can be accommodated mathematically. Much is now known about planetary motion, and space is 'empty' with few extraneous factors to take into account, so the planet's motion can be described very accurately. This is essential, of course, if the model is to be used predictively (e.g. if a space agency is planning a rocket flight). The tank of still water is also fairly easily described, but now there are some awkward factors (such as fluid viscosity, laboratory temperature, smoothness of stone and so on) that might have an influence, so the accuracy of the derived equations is likely to be lower in this case. However, if this model is simply a descriptive one then accuracy is less of a consideration than with the planetary motion. Finally, with the pebble on the beach there will be a myriad influencing factors such as atmospheric conditions, height of tide, presence of rocks, velocity of projection etc. These may be difficult factors to treat mathematically, so any derived equations can at best be viewed as approximations only. Nevertheless, even crude approximations can be useful as a starting point for practical progress.

There is one central feature of these mathematical models that is of considerable practical significance. This concerns the handling of imposed constraints and external influences. One way of proceeding is to assume such factors to be completely specifiable mathematically, which implies that they are exactly predictable and not affected in any way by chance fluctuations. This is a perfectly satisfactory assumption to make in many circumstances, especially those situations in which chance plays a negligible part (for example planetary motion, or reactions of chemicals) or where the objective is merely to obtain a broad *overall* description of the system. However, for anything that requires more *individual* precision it may not be acceptable, since chance plays a major role in most situations in life and any numerical data collected as part of an investigation will be subject to many chance fluctuations (sometimes dramatically so). Consider, for example, a pharmaceutical company manufacturing a new headache tablet. The company sets up a clinical trial to establish whether this tablet is more efficacious than one manufactured by a rival company. However, not only might the two tablets react differently to different severities of headache, but different people in the trial will almost certainly differ wildly in their tolerances, susceptibilities or reactions to the tablets. All such chance fluctuations must be taken into consideration if correct conclusions are to be drawn from the results observed in the trial. Similarly, any opinion poll or market research investigation must take account of the notorious fickleness in people's declared preferences: any one person might give two very different answers to the same question at different times and this degree of 'error' must be accommodated in any analysis of resulting data. Likewise, biological organisms can exhibit large variation in the size and shape of otherwise 'similar' entities, so any measurements made on these organisms will also exhibit large variation which must be allowed for in any analysis. Such variation is pervasive in most substantive applications that involve the collection of quantitative information.

Valid analysis of either observational or experimental data must therefore be founded on techniques that pay due regard to possible chance fluctuations. By incorporating chance into our equations in some way, we turn mathematical models into *statistical* models, and it is with such models that this book is concerned. A statistical model provides the bedrock on which most techniques of data analysis are founded, and the aim of this book is to provide the student with a comprehensive account of those models that are most useful in the analysis of typical forms of data as encountered in a wide range of practical situations. We will build up the logical basis of some general classes of models, formulate principles of data analysis based on such models, and develop a number of specific models together with methods of analysing data based on them. However, first we must establish the necessary foundation. The general background ideas and terminology occupy the remainder of this chapter, and the essential background theory of probability and basic inference is reviewed in Chapter 2. The main development then commences in Chapter 3.

1.2 Populations and samples

The essence of any investigation is the gathering of information. Statistics is brought into play whenever the information gathered is *quantitative* and pertains to a *collection of individuals*. In this context, 'individuals' can be either animate or inanimate (e.g. students at a particular university, houses in a particular area of a city, bales of hay harvested on a particular farm) while the most common types of information are either measurements or counts of a particular feature of each individual (e.g. height of each student, floor area of each house, number of rooms in each house, dry weight of each bale). Commonly, individuals are termed *units* while the measured features are termed *variables*. Sometimes, a variable of interest might be *qualitative* with a number of distinct possible *states* (e.g. 'colour of eyes' with states blue, green, brown etc). The usual form of data for such variables is as counts that show the numbers of individuals in each of these states (e.g. number of blue-eyed students, number of brown-eyed students etc).

Most investigators wish to make statements about *populations*. A population is the complete set of individuals that could be measured. Sometimes this set will be finite (e.g. all students at Exeter University in 1997, all houses in Plymouth), and sometimes it will be infinite (e.g. all bales of hay that might ever be harvested in Devon). If a population is finite, then in principle every individual in it could be examined and the value of any variable of interest could be measured on it. Thus a complete description of the population could be obtained. However, in most practical situations such examination would take up so much time and expense as to be completely ruled out. Moreover, if a population is infinite then it is not possible to examine every individual in it. Thus in nearly all practical investigations, the data for analysis will come from only a subset of the population, i.e. a *sample* from it. The hope is that the sample will be *representative* of the whole population, so that features of the population will be mirrored by those of the sample. Then the investigator will seek to draw *inferences* about the population from analysis of the sample.

To ensure that a sample is truly representative of the population from which it comes, it is essential to guard against any systematic bias in the choice of individuals to include in the sample. There are in general many considerations to be taken into account when deciding how to draw a sample, but such considerations lie outside the scope of this book. It here suffices just to say that in most practical circumstances the simple process of *random sampling* is enough to ensure not only that the sample is representative of the population but also that most of the assumptions we will make later when constructing statistical models are reasonable ones. A random sample is one in which every member of the population has an equal chance of being included. If the population can be listed (e.g. student enrolment register, housing list, electoral roll, etc), then a random sample of size *n* is obtained by choosing *n* numbers off the list 'at random' and including the corresponding individuals in the sample. Modern computer packages such as MINITAB contain pseudo-random number generators which enable this choice to be made objectively so as to satisfy the necessary probability condition. Alternatively, recourse can be made either to tables of random numbers (as given in most books of statistical tables, such as that by Murdoch and Barnes, 1970) or to some randomization device such as dice or cards to achieve the same ends.

Full details of the process of sampling may be found in a specialist text, such as *Elements of Sampling Theory* by V. Barnett (English Universities Press, 1974). For our purposes we will assume that the sampling has already been conducted in an appropriate manner, so that any samples we encounter are representative of the population they have come from, are small in size relative to that of the population, and have been chosen so as to satisfy the criterion of randomness.

1.3 Variables and factors

Having considered the selection of individuals in a study, we now turn to a closer examination of the observations made on each of these individuals. We have already mentioned above that the measured features are usually termed *variables*, and we have distinguished between quantitative (i.e. numerical) and qualitative (i.e. non-numerical) variables. Sometimes, just a single variable is measured on each individual and the purpose of the study is to characterize the values of this variable in a specified target population. For example, a local authority might be reviewing its provision of travel grants to students within its jurisdiction, so might wish to obtain information on the population of distances travelled by students between their homes and their places of study. Distance travelled is thus the sole variable of interest, the objective perhaps being to determine a minimum value for grant eligibility that will ensure the authority's budgeted funds are sufficient to meet the demand.

More commonly, however, several variables will have been measured on each individual, either because of their intrinsic interest to the investigator, or because it is convenient and efficient to collect as much information as possible once a (perhaps costly) investigation has been set up. Moreover, these variables can generally be put into one of two categories: those of primary interest to the investigator (i.e. those which he or she wishes to study explicitly) and those which might provide supplementary or background information. Variables of the former type are called *response* variables, while those of the latter type are called *explanatory* variables. Alternative names commonly used for 'response' and 'explanatory' are *dependent* and *independent* respectively. Unfortunately, while the word 'dependent' carries unambiguous connotation of dependence on other variables, 'independent' might be taken to imply that the explanatory variables are *statistically* independent of each other (which is most definitely *not* the case in most practical situations). Use of the word 'independent' might therefore lead to confusion. We shall see later that *regressor* is a suitable alternative name for such variables, but till then will keep to the term 'explanatory'.

As an example, health workers might be interested in studying the incidence of respiratory problems in a particular population of individuals, and might conduct a survey asking respondents to state the number of visits they have made to the doctor in the previous year with respiratory complaints. While the results of the survey might provide useful information in themselves, they will clearly be enhanced and made much more specific if information on the respondents' life styles is collected at the same time. Suitable explanatory variables that would be expected to have some bearing on respiratory ailments might be age and occupation of respondent, number of cigarettes smoked each week, amount of alcohol drunk each week, and so on. Relating the data on surgery visits to these extra variables will help to disentangle the various factors responsible for respiratory ailments.

One of the main areas of application of statistical models is in the study of dependence between response and explanatory variables. Construction of such models requires all variables to be numerical, so that mathematical equations describing the relationship can be formulated. This presents no problem in the case of quantitative variables such as age, number of cigarettes smoked or amount of alcohol drunk (if this is expressed in a quantitative measure such as number of units) in the example above, but what about a qualitative variable such as occupation (which might have states 'manual', 'clerical', 'professional', etc)? A similar question arises when a variable is quantitative but its value is only observed in categorical fashion (e.g. none, less than 10, between 10 and 20, more than 20, as four possible 'values' for the variable 'number of cigarettes smoked'). The answer to this question is to introduce one or more new variables, which define in a unique and numerical fashion the state of a qualitative variable. Such variables are termed *dummy variables* (because they are not explicitly observed, but have their values *assigned* to them by the analyst).

The simplest qualitative variable is one with just two possible states (e.g. present/absent, yes/no, male/female, etc), and an appropriate dummy variable would be formed by assigning values 0 and 1 (say) to the different states. In general, if a qualitative variable has m possible states then we need $m - 1$ such dummy variables, each again taking value 0 or 1, to represent it. Labelling the m states (arbitrarily) 1 to m and the dummy variables x_1 to x_{m-1}, one possible assignment is to set x_i equal to 1 and all other xs equal to zero whenever the individual is in state i for $i = 1, 2, \ldots, m - 1$, and all xs equal to zero whenever the individual is in state m. Other (equivalent) definitions are obtained by permuting either states or dummy variables.

Thus, for example, if 'occupation' above has the states 'manual', 'clerical', 'professional' and 'other', then we would need to define three dummy variables x_1, x_2 and x_3. A respondent whose state was 'manual' would be assigned the values $x_1 = 1$, $x_2 = 0$, $x_3 = 0$, one whose state was 'clerical' the values $x_1 = 0$, $x_2 = 1$, $x_3 = 0$, one whose state was 'professional' the values $x_1 = 0$, $x_2 = 0$, $x_3 = 1$, and one whose state was 'other' the values $x_1 = 0$, $x_2 = 0$, $x_3 = 0$.

Variables that have just two possible values are called *binary* or *dichotomous* variables. Note also that values other than 0/1 can be used, but in general the simplest choice is 0/1.

A final point of terminology concerns the word *factor*, which is used in several ways. In the broad sense, *any* general explanatory feature is referred to as a factor; for example, 'age', 'occupation', 'smoking' and 'alcohol' are all factors that might have an influence on respiratory health. However, in a narrower technical sense the word 'factor' is usually synonymous with *qualitative explanatory variable*, whose states are then called *levels* of the factor. Thus 'occupation' above is a factor, with the four levels 'manual', 'clerical', 'professional' and 'other', but if 'age' is expressed in years then it is still referred to as a variable rather than as a factor.

1.4 Observational and experimental data

Data may be collected in a variety of ways, but a fundamental dichotomy that is useful in statistics is one between an *experiment* and a *survey*. These two words conjure up specific images in most people's minds: a white-coated individual in a laboratory surrounded by complicated equipment on the one hand, and an earnest questioner with a clipboard accosting people in the street and asking them many questions on the other. These images are of course appropriate, but they represent only a small part of all the situations that can be grouped into the two classes defined by the terms.

In statistics, a *survey* can be used to describe any occasion on which data have been collected simply by observation. The survey is thus typified by random sampling of individuals from a large target population of such individuals, followed by measurement of chosen variables on each individual. The defining characteristics here are that (i) *the data collection takes place in 'real' surroundings*, and (ii) *the researcher has no control over prevailing conditions*. Thus an agriculturalist wishing to investigate relationships between height of plants on the one hand and ambient temperature and soil moisture on the other might simply go out to a number of places and measure the three variables on randomly chosen plants in each place. There is nothing wrong with this procedure, but on returning home and examining the data the investigator may find that the range of ambient temperatures and/or soil moistures is so small as to make it difficult to establish meaningful relationships from the sample data.

A better approach in such an application might be for the agriculturalist to *grow* some plants in a greenhouse, say, where the conditions can be controlled so that a *range* of both soil moistures and ambient temperatures can be produced. Then measurements of heights of plants at *different* soil moistures and ambient temperatures are more likely to enable relationships to be established. This is an example of an *experiment*, the defining features of which are the opposite of those of a survey: (i) *the data collection takes place in artificial surroundings*, and (ii) *the researcher has complete control over prevailing conditions*.

Of course there might be drawbacks with experimental data, for example in that results might not be easily extrapolated from the artificial surroundings to the 'real world' in which we want to use them. However, it is important to recognize the two distinct situations, as we might wish to take a slightly different viewpoint when analysing the resulting data in each. Where necessary, in succeeding chapters we will draw attention to the type of situation for which a particular modelling technique is appropriate.

1.5 Statistical models

We have now arrived at the stage where an investigator has drawn a sample of individuals from a target population, and has measured a set of variables on each sample individual. The investigator will typically also have some objective in mind when collecting the data. Most often this will arise from some theory, or hypothesis, that he or she wants to test, typically about the *general* behaviour of the system under study. For example, this theory might concern the relationship among the measured variables, or the difference in response on average between two specified populations of individuals, or (more specifically) the effects of different combinations of fertilizers on the harvested grain in different localities. A necessary first step in any analysis of the data is to specify a mathematical expression for this general behaviour that embodies the experimenter's beliefs in an appropriate fashion; this expression is known as the *systematic component* of the model.

However, as we are aware from above, all individuals exhibit a certain amount of idiosyncratic behaviour, and the values recorded for a sample of individuals on any variable will show a certain amount of fluctuation. This fluctuation will be more or less pronounced according to the variable being measured. For example, whereas human pulse rate varies markedly from person to person, body temperature is much more consistent among healthy subjects. The general behaviour, i.e. systematic model component, referred to above is intended to describe an 'ideal' response of an individual, so to take account of the known fluctuations in responses we must include a *random component* in the model and also specify how the random and systematic components are *combined*. The formulation of the systematic component, the assumptions made about the random component and the specification of how the two components are combined constitute the three essential features of a statistical model.

We will be considering these three features of a model in considerable detail in subsequent chapters, but a few general introductory points are in order here. First, the systematic component. This often expresses some population feature of the response variable (e.g. its expected value) in terms of the explanatory variables through a *parametric equation*. For example, suppose that the response variable is y, there are two explanatory variables x_1 and x_2, and the expected value of y (denoted μ) is supposed to depend on these explanatory variables in some way. Three possible relationships might be the following:

1. $\mu = \beta_0 + \beta_1 x_1 + \beta_2 x_2$;
2. $\mu = \beta_0 \exp(\beta_1 x_1 + \beta_2 x_2)$;
3. $\mu = \beta_0 + \beta_1 x_1^2 + \beta_2 x_2^2$.

Each of these equations specifies a particular *form* of relationship, while the quantities $\beta_0, \beta_1, \beta_2$ are *parameters* that can be varied to provide different specific relationships for each form. The parameter β_0 is a 'baseline' *constant* in each case, giving the value of μ when x_1 and x_2 are both equal to zero, while the parameters β_1 and β_2 are the *coefficients* of x_1 and x_2 respectively, indicating the influence that each of these variables has on the value of μ. One of the main objectives of modelling is to determine the 'best' values to assign to such parameters for any particular set of data. Methods of estimating such 'best' values will be discussed later, but one crucial consideration in developing these methods is the *type* of relationship being proposed. Here, a fundamental distinction has to be made between *linear* and *non-linear* models.

When we speak of linearity or non-linearity of models it is to the *parameters* of the model that we attach these descriptors. Model 1 above is a *linear* model in this sense. The defining feature of a linear model is that a unit change in any parameter value leads to the same change in the dependent variable *whatever the values of the parameters*. Thus in model 1, suppose that the values of β_0, β_1 and β_2 are a, b, c respectively. Then $\mu = a + bx_1 + cx_2$. Now suppose that β_1 changes to $b+1$. Then the change in μ is $[a+(b+1)x_1+cx_2]-[a+bx_1+cx_2] = x_1$, and this will be the same *whatever the values of a, b and c*.

Applying the same manipulations to model 3, we see that the change in μ this time is x_1^2 which also remains the same on varying a, b or c. Thus model 3 is also a linear model. It should therefore be noted that any non-linearity of *variables* in a model (here the squares x_1^2 and x_2^2) is irrelevant; it is in terms of the parameters that the linearity property has to be tested.

However, this property does not hold in model 2. Taking the same values as above, at $\beta_0 = a$, $\beta_1 = b$, $\beta_2 = c$ we have $\mu = a\exp(bx_1 + cx_2)$. When β_1 changes to $b + 1$, the change in μ is $a\exp([b + 1]x_1 + cx_2) - a\exp(bx_1 + cx_2)$, which can be simplified to $a(e^{x_1} - 1)\exp(bx_1 + cx_2)$ but no further. Thus, this change in μ will be *different for different values of a, b and c*. Hence model 2 is a *non-linear* model. Note, however, that if we take logarithms of both sides of the equation in model 2 we obtain $\log\mu = \log\beta_0 + \beta_1 x_1 + \beta_2 x_2$. This is of precisely the same form as model 1, but with the new parameters $\log\mu$ and $\log\beta_0$ in place of μ and β_0. Thus the non-linear model 2 can be *reparameterized* to turn it into a linear model. Some non-linear models can be *linearized* in this fashion, but other non-linear models can never be linearized. An example of an *intrinsically non-linear* model, i.e. one that cannot be linearized, is $\mu = \beta_0\exp(\beta_1 x_1)+\exp(\beta_2 x_2)$. In general, the handling of non-linear models is much more complicated than the handling of linear ones, so it is on the latter that we will principally focus in this book.

If the systematic component of a model is concerned with overall population features such as expected values, the random component focusses on the individual fluctuations in a system. To account for individual unpredictability we need to turn to a *probabilistic* formulation, and the natural way to describe sampling behaviour is through a *probability distribution*. There are many probability distributions that could be used as models for a variable in any given situation, the appropriate choice being guided by the sort of variable that is observed. For example, a normal distribution would be a natural choice for measurement data, a Poisson model for count data, and a binomial model for proportions. We review some of these common models and their properties in Chapter 2.

The final ingredient in the modelling process is the method of combining the systematic and random components. There are various possibilities, but by far the most common method in practice is to combine them *additively*. A very simple example will illustrate the reasoning behind this choice. Suppose that an experiment is conducted on the effect that two different chemicals have on the growth of a fungus on leaves. Different leaves are treated with varying concentrations (x_1, x_2) of the chemicals, and the dry weight (y) of fungus on each leaf is found after a fixed period of growth. It is reasonable to assume that each weighing will be an observation from a normal distribution whose mean will depend on the concentrations of chemicals in some way, but whose variance will be constant over all weighings. If this constant but unknown variance is denoted by σ^2, and if we assume the linear model 1 above for the dependence of the population mean on the concentrations of chemicals, then a weight of fungus y_i which was subjected to chemical concentrations x_{1i}, x_{2i} will be an observation from a normal distribution that has mean $\beta_0 + \beta_1 x_{1i} + \beta_2 x_{2i}$ and variance σ^2. This can be re-expressed by saying that $y_i = \mu_i + \epsilon_i$, where $\mu_i = \beta_0 + \beta_1 x_{1i} + \beta_2 x_{2i}$ and ϵ_i is an observation from a normal distribution with mean zero and variance σ^2 [symbolically written $\epsilon_i \sim N(0, \sigma^2)$]. In this expression, μ_i is the systematic component, ϵ_i is the random component, and the two are combined additively. This is a very simple example of the type of model that we will be concerned with in this book.

As a final point it is worth noting that sometimes the systematic and random components are combined multiplicatively rather than additively. In certain circumstances such models can be reduced to additive ones on the logarithmic scale (in the same manner that some non-linear models can be linearized by taking logarithms). However, we will not pursue either non-linear or multiplicative models to any extent in this book, but will concentrate in the main on linear additive ones.

2
Distributions and inference

For a good understanding of the material in this book, the reader should be thoroughly familiar with those basic ideas and results of probability, random variables, probability distributions and statistical inference that are typically covered in a university first-year introductory course in probability and statistics. While this level of knowledge is therefore assumed, it is nevertheless convenient to collect together here in summary form some relevant facts and formulae for later reference. Those readers whose statistical foundations are either sketchy or rusty will benefit from careful study of this material, but others should find a quick skim through the chapter to be sufficient both to refresh the memory and to set the results in the modelling context. In particular, the very last section introduces some material more advanced than that usually found in a first course, and this material will be taken up in a later chapter.

2.1 Random variables and probability distributions

We have seen in the introductory chapter that attention is often focussed in statistical work on some particular response variable, and that values of this variable are generally collected from a set of individuals. A defining feature of all response variables of interest in practice is that it is impossible to predict in advance of observation what the actual value of the variable will be for any specific sample individual. We will typically know that the value comes from some range of values, and that some values within this range are more likely to occur than others, but the exact value will only be known *after* observation or measurement has been made. In view of this inherent unpredictability, such a variable is said to be a *random variable*.

Symbolically it is convenient to denote a random variable by an upper-case letter, say Y, and any value observed for it by a lower-case version of the same letter, subscripted if necessary to distinguish separate values. Thus y_1 will denote the value of Y for the first individual observed, y_2 the value for the second individual observed, and so on. A sample of n values will be written $y_1, y_2, \ldots, y_n$.

For technical reasons that will become obvious, a distinction has to be made between *discrete* and *continuous* random variables. A discrete variable is one whose possible values come from either a finite or a countably infinite set, so that these values can all be listed in a (possibly infinite) sequence. A continuous variable, on the other hand, is one whose values can be any real numbers in a specified range. Such a set of values is uncountably infinite (since it can be shown that between any two given real numbers there always exists another real number), so it is not possible to list all values of a continuous variable in a sequence.

A complete description of a random variable is given by specifying all the values it could conceivably take and quoting their associated probabilities; such a specification is known as the *probability distribution* of the variable. In the case of a discrete random variable, where all possible values can be listed in sequence, there is no problem about drawing up its probability distribution. For example, suppose that Y denotes the sum of the values on the upper faces of two randomly thrown dice. Each die has possible values 1, 2, 3, 4, 5, 6, so Y can take any integral value between 2 and 12 inclusive. There are 36 distinguishable and equally likely outcomes from throwing two dice, so to obtain the probability that Y takes a specified value y we simply need to find how many of these outcomes yield $Y = y$ and divide this number by 36. Doing this for all y leads to the probability distribution:

y	2	3	4	5	6	7	8	9	10	11	12
$\Pr(Y = y)$	$\frac{1}{36}$	$\frac{2}{36}$	$\frac{3}{36}$	$\frac{4}{36}$	$\frac{5}{36}$	$\frac{6}{36}$	$\frac{5}{36}$	$\frac{4}{36}$	$\frac{3}{36}$	$\frac{2}{36}$	$\frac{1}{36}$

While such a display is perfectly acceptable, it becomes unwieldy once there are more than just a few possible values of the random variable (and, of course, impossible if this number is infinite!). A much more compact and universal presentation is provided by a formula, from which the probability of any observed value of the random variable can be worked out, together with the set of possible values of the variable. The above probability distribution can thus be more compactly expressed as:

$$\Pr(Y = y) = \frac{1}{36} \times \min(y - 1, 13 - y) \qquad \text{for } y = 2, 3, \dots, 12 \qquad (2.1)$$

We often write $f(y) = \Pr(Y = y)$, and refer to $f(y)$ as the *probability density function* of Y. (Strictly speaking we should call it the probability *mass* function, but the use of the word 'density' will ensure the same terminology for both discrete and continuous variables). Clearly, from basic properties of probability, the density function satisfies the two conditions:

(i) $f(y) \geq 0$ for all y (as no probability can be negative),

(ii) $\sum f(y) = 1$, where the sum is taken over all possible values of Y (as one of the outcomes *must* occur at any given trial).

Another important quantity is the *distribution function* of a random variable. This function is defined as

$$F(y) = \Pr(Y \le y), \tag{2.2}$$

so if Y is a discrete variable then $F(y) = \sum_{r=-\infty}^{y} f(r)$. It is evident from this definition that the distribution function satisfies the following properties:

(i) $F(-\infty) = 0$;
(ii) $F(\infty) = 1$;
(iii) if $a \le b$ then $F(a) \le F(b)$, i.e. F is a non-decreasing function.

Turning to continuous random variables, it is most convenient to work from the distribution function as defined in (2.2). If we know all the probabilities of the form $\Pr(Y \le y)$, then we can define the probability that Y lies in any interval:

$$\Pr(a < Y \le b) = \Pr(Y \le b) - \Pr(Y \le a) = F(b) - F(a).$$

Corresponding to every continuous random variable Y is a function $f(y)$ defined by

$$f(y) = \frac{d}{dy} F(y),$$

that is

$$F(y) = \int_{-\infty}^{y} f(t) \, dt.$$

It follows from above that $f(y)$ satisfies the conditions

(i) $f(y) \ge 0$ for all y,
(ii) $\int_{-\infty}^{+\infty} f(y) \, dy = 1$,

and $\Pr(a < Y \le b) = \int_{a}^{b} f(y) \, dy$. This function $f(y)$ is again called the probability density function of Y.

Interpretation of the probability density function requires just a little care. By taking $a = c - \frac{1}{2}\delta$ and $b = c + \frac{1}{2}\delta$ for any real number c and *small* δ, we have

$$\Pr\left(c - \frac{1}{2}\delta < Y \le c + \frac{1}{2}\delta\right) = F\left(c + \frac{1}{2}\delta\right) - F\left(c - \frac{1}{2}\delta\right).$$

If we then apply a first-order Taylor expansion to each of these distribution functions and use the definition of the density function as the derivative of the distribution function, we find

$$\Pr\left(c - \frac{1}{2}\delta < Y \le c + \frac{1}{2}\delta\right) \approx f(c)\delta.$$

The probability density function $f(y)$ is thus proportional to the probability that Y lies in a small interval centred on y.

A common misconception is to view the probability density function $f(y)$ as the probability that Y equals y, but a simple argument shows that this is false. First note that $\Pr(a < Y \leq b)$ is always greater than $\Pr(Y = b)$. Letting a tend to b we see that the above integral for the first probability approaches zero, and we therefore conclude that $\Pr(Y = b) = 0$ for any single real number b.

At first sight this result would appear to negate the usefulness of working with continuous random variables, as it seems to suggest that any particular value of Y is impossible. However, in practice this is not a problem, because measuring random variables to a finite number of decimal places effectively means that we are always dealing with intervals. It also means that we do not need to distinguish between inclusive inequalities and strict inequalities when dealing with continuous variables, i.e.

$$\Pr(a \leq Y \leq b) = \Pr(a < Y < b).$$

2.2 Probability distributions as models

We have seen in Chapter 1 that the main purpose in most statistical investigations is to obtain information about a population of individuals from observations made on a sample of its members. However, when we talk about a population it is not the individuals themselves that concern us, but the variable or variables that we are measuring on them. For example, suppose that we are interested in obtaining information on the distances travelled between their home town and Exeter by first-year students at Exeter University. The population of targeted individuals is 'all first-year students at Exeter University', but of course each of these individuals can supply information on very many different response variables (such as height, weight, number of A level examinations passed, and so on). Since we are interested in just one specific such variable, namely $Y = $ 'distance travelled from home to Exeter', then the relevant population is the set of all possible y *values* for first-year students at Exeter.

Thus all populations will henceforth be taken to be populations of values of some specified variable(s). If the population is infinite then we can never obtain all its values, and even if the population is finite we do not usually want to measure all its values. In either case, we will wish to obtain information on particular *features* of the population from a restricted number of sample values. For example, we might want to know what distance is travelled *on average* by first-year students from home to Exeter, or perhaps the distance within which seventy-five percent of all first-year students live. To proceed we need to formulate a suitable *model* for the y values that make up the population, relate the features of interest to appropriate aspects of the model, and then use the sample data to provide estimates of these aspects.

The essential feature of such y values is their unpredictability, and the best we can do is to specify the probability of obtaining either a given value or a value in a given range. Thus probability distributions provide the most appropriate models for response variable populations. More specifically, either the probability density function $f(y)$ or the probability distribution function $F(y)$ provides the necessary information so constitutes the population model for Y. Of course, since we never actually *know* this distribution we have to assume some general functional form for either $f(y)$ or $F(y)$. This functional form usually involves one or more *parameters* that can be varied, and the hope is that there will be some specific values of these parameters for which the resulting distribution *fits* our observed data adequately. Such a model is called a *parametric model* for Y.

We will consider choice of such a functional form in the next section. For the present let us suppose that we have specified some suitable function $f(y)$ as the probability density of our population. What summary values of this density might relate to those population features that are usually of interest? To answer this question, let us focus on discrete response variables and interpret probability as 'long-run relative frequency'. Thus if Y has possible values $y_1, y_2, y_3, \ldots$ and if $\Pr(Y = y_i)$ is denoted p_i then we can think of p_i as being the relative frequency with which the value y_i occurs in the whole population. We define the *expected value* of Y by the equation:

$$E(Y) = \sum_i y_i p_i. \tag{2.3}$$

Using the relative frequency interpretation of p_i given above, $E(Y)$ can be interpreted as the average of the y values across the whole population so it is the *population mean*. This is clearly one of the most important summary values for the population. It is often denoted by μ and since it measures the 'centre' of the population it is also known as the *location parameter* of the population.

Although the variable of interest is Y, we may frequently want to consider some *function* $g(Y)$ of it. Then if the values of Y are $y_1, y_2, \ldots$, the values of the function will be $g(y_1), g(y_2), \ldots$. (A trivial example might be if we have measured heights in inches, but then want to work in centimetres. In this case we have started with $Y =$ height in inches, but we subsequently consider $g(Y) = 2.54 \times Y =$ height in centimetres.) Since p_i gives the probability of observing $Y = y_i$, it will obviously also give the probability that $g(Y) = g(y_i)$. The population average, i.e. *expected value* of $g(Y)$ is thus:

$$E\{g(Y)\} = \sum_i g(y_i) p_i. \tag{2.4}$$

One function of particular interest in this context is $g(Y) = Y^r$, i.e. the rth power of Y, and using this function in the above equation yields the rth *(central) moment* of Y:

$$E(Y^r) = \sum_i y_i^r p_i. \tag{2.5}$$

Another function of interest is $g(Y) = (Y - \mu)^r$, which gives the rth power of the difference between Y and the population mean, and using this function in the equation gives the rth *moment about the mean* of Y:

$$E\{(Y - \mu)^r\} = \sum_i (y_i - \mu)^r p_i. \tag{2.6}$$

Using the relative frequency interpretation of probability, each of these moments can be viewed as the average of the relevant function across all the y values in the population. When $r = 1$, the first central moment is just $E(Y)$, which we have already seen to be the population mean, while the first moment about the mean $E\{(Y - \mu)\}$ is the average of the differences between each y_i and the population mean. However, since the population mean is itself the average of the y_i it is evident that the values $y_i - \mu$ must be equally balanced about zero, some negative and some positive, so that the first moment about the mean must always be zero. On the other hand, when $r = 2$ the second moment about the mean yields the average of the squared differences between each y_i and the population mean so measures the *spread* of the values about μ. The process of squaring removes all negative values, so the second moment about the mean can never be negative. It is called the *variance* of the population, and is often denoted by σ^2:

$$\sigma^2 = E\{(Y - \mu)^2\} = \sum_i (y_i - \mu)^2 p_i. \tag{2.7}$$

A little bit of simple algebra establishes the identity

$$E\{(Y - \mu)^2\} = E(Y^2) - \mu^2, \tag{2.8}$$

the latter usually being easier to work with. The square root of the variance, σ, is the *standard deviation*. The quantity σ^2 is often referred to as the *dispersion* parameter of the population.

A large standard deviation or variance indicates a distribution whose values are widely spread about the mean, while a small standard deviation or variance indicates one whose values are tightly bunched about the mean. Of course, the absolute size of the standard deviation or variance depends on the actual values taken on by the variable, so 'large' or 'small' is a relative concept (e.g. if every value of Y is multiplied by 10, then the standard deviation is multiplied by 10 and the variance is multiplied by 100). A dimensionless measure of variability that is sometimes employed is the *coefficient of variation*, defined to be the standard deviation divided by the mean and multiplied by 100 to express it as a percentage. This coefficient is unaffected by any scaling of values of Y, but it does depend heavily on the mean and makes most sense when variables can take only positive values.

The location and dispersion are the two population features of most interest in practice, but occasionally higher moments might also be of interest. The *skewness* γ of the distribution involves the third moment about the mean:

$$\gamma = E\{(Y - \mu)^3\} \div \sigma^3, \tag{2.9}$$

while the *kurtosis* κ of the distribution involves the fourth moment about the mean:

$$\kappa = E\{(Y - \mu)^4\} \div \sigma^4. \tag{2.10}$$

Skewness measures the extent of asymmetry of the distribution: a symmetrical distribution has zero skewness, a distribution with a long 'tail' of high values but a bunching of low values has positive skewness, while one with a long 'tail' of low values but a bunching of high values has negative skewness. Kurtosis measures the 'flatness' of a distribution: one with a large kurtosis has a 'flat' distribution (i.e. many values y_i whose corresponding p_i are comparable and appreciable) while one with a small kurtosis has a 'peaked' distribution (i.e. a few y_i with very large p_i, the rest with very small p_i). The normal distribution (see below) is often used as a 'standard', and it has a κ value of 3. The divisors in their definitions ensure that both skewness and kurtosis are dimensionless quantities, unaffected by either additive or multiplicative transformations of Y.

Example 2.1. To illustrate these ideas, consider the distribution given above for Y = sum of values on the upper faces of two randomly thrown dice.

First, $\mu = \frac{1}{36} \times 2 + \frac{2}{36} \times 3 + \frac{3}{36} \times 4 + \cdots + \frac{2}{36} \times 11 + \frac{1}{36} \times 12 = 7$. The population mean is thus 7; this is the value we would obtain as the average of very many throws of two dice, and hence is the value we 'expect' to obtain in a single throw.

Thus $E\{(Y-\mu)^2\} = \frac{1}{36} \times (2-7)^2 + \frac{2}{36} \times (3-7)^2 + \cdots + \frac{2}{36} \times (11-7)^2 + \frac{1}{36} \times (12-7)^2 = 5.8333$. (Note also that $E(Y^2) = \frac{1}{36} \times 2^2 + \frac{2}{36} \times 3^2 + \cdots + \frac{2}{36} \times 11^2 + \frac{1}{36} \times 12^2 = 54.8333$ so that $E(Y^2) - \mu^2 = 54.8333 - 49 = 5.8333 = E\{(Y-\mu)^2\}$.) Thus the variance σ^2 of the distribution is 5.8333, so the standard deviation σ is $\sqrt{5.8333} = 2.415$ and the coefficient of variation is $100 \times 2.415 \div 7 = 34.5\%$.

Next $E\{(Y - \mu)^3\} = \frac{1}{36} \times (2 - 7)^3 + \frac{2}{36} \times (3 - 7)^3 + \cdots + \frac{2}{36} \times (11 - 7)^3 + \frac{1}{36} \times (12 - 7)^3 = \frac{1}{36} \times (-125) + \frac{2}{36} \times (-64) + \cdots + \frac{2}{36} \times (64) + \frac{1}{36} \times (125) = 0$ (the positive and negative contributions exactly balancing from the middle of the distribution outwards). Thus the skewness $\gamma = 0 \div (2.415)^3 = 0$ and the distribution is symmetric (evidently so).

Finally, $E\{(Y - \mu)^4\} = \frac{1}{36} \times (2 - 7)^4 + \frac{2}{36} \times (3 - 7)^4 + \cdots + \frac{2}{36} \times (11 - 7)^4 + \frac{1}{36} \times (12 - 7)^4 = \frac{1}{36} \times (625) + \frac{2}{36} \times (256) + \cdots + \frac{2}{36} \times (256) + \frac{1}{36} \times (625) = 80.5$ so that the kurtosis is $\kappa = 80.5 \div 2.415^4 = 80.5 \div 34.027 = 2.366$. The shape of the distribution overall is thus very close to, but slightly more peaked than, the normal distribution.

We can carry all the above ideas across to continuous variables; to do so we simply need to replace individual probabilities by probability densities and summation by integration. Assuming that Y is now a continuous random variable defined on a range R of values and with probability density function $f(y)$, the main results are as follows.

The expected value of Y is given by

$$E(Y) = \mu = \int_R yf(y)\,dy. \tag{2.11}$$

Likewise, the expected value of the function $g(Y)$ is

$$E\{g(Y)\} = \int_R g(y)f(y)\,dy, \tag{2.12}$$

and replacing $g(Y)$ in turn by Y^r and $(Y - \mu)^r$ yields the rth central moment

$$E(Y^r) = \int_R y^r f(y)\,dy \tag{2.13}$$

and the rth moment about the mean

$$E\{(Y - \mu)^r\} = \int_R (y - \mu)^r f(y)\,dy. \tag{2.14}$$

Using the above expectations in analogous fashion to before, we arrive at the following population quantities:

(i) mean $\mu = \int_R yf(y)\,dy$;
(ii) variance $\sigma^2 = E\{(Y - \mu)^2\} = \int_R (y - \mu)^2 f(y)\,dy$;
(iii) standard deviation $\sigma = \sqrt{\text{variance}}$;
(iv) coefficient of variation $= \frac{\sigma}{\mu} \times 100\%$;
(v) skewness $\gamma = E\{(Y - \mu)^3\} \div \sigma^3$; and
(vi) kurtosis $\kappa = E\{(Y - \mu)^4\} \div \sigma^4$.

As before, it is easy to show that $E\{(Y - \mu)^2\} = E(Y^2) - \mu^2$ which helps with calculations. Although these definitions now involve a number of mathematical subtleties, for practical purposes all interpretations of the various quantities are the same as before.

It frequently happens that we have to deal with two or more random variables simultaneously. If Y_1 and Y_2 are two (discrete) random variables defined on the same sample space, then the function

$$f(y_1, y_2) = \Pr(Y_1 = y_1 \text{ and } Y_2 = y_2)$$

is called their *joint probability density function*. This function may be given either by a formula or in tabular form, but in either case satisfies the condition $\sum_{y_1} \sum_{y_2} f(y_1, y_2) = 1$, where the summations are over all possible values of y_1 and y_2.

We can calculate distributions of the individual random variables from this joint distribution. Since $\Pr(Y_1 = y_1) = \sum_{y_2} \Pr(Y_1 = y_1 \text{ and } Y_2 = y_2)$, it follows that the distribution of Y_1 is

$$f_{Y_1}(y_1) = \sum_{y_2} f(y_1, y_2).$$

This is called the *marginal distribution* of Y_1. A corresponding relationship holds for the marginal distribution $f_{Y_2}(y_2)$ of Y_2. The random variables Y_1 and Y_2 are *independent* if $f(y_1, y_2) = f_{Y_1}(y_1)f_{Y_2}(y_2)$ at all points y_1, y_2.

We define expectation and moments of joint distributions in the obvious way. Thus for any function $g(y_1, y_2)$ we define

$$E\{g(Y_1, Y_2)\} = \sum_{y_1} \sum_{y_2} g(y_1, y_2)f(y_1, y_2).$$

In particular we have $E(Y_1 Y_2) = \sum_{y_1} \sum_{y_2} y_1 y_2 f(y_1, y_2)$, from which we define the *covariance* between Y_1 and Y_2 as

$$\mathrm{cov}(Y_1, Y_2) = E(Y_1 Y_2) - E(Y_1)E(Y_2)$$

and the *correlation* between Y_1 and Y_2 as

$$\mathrm{corr}(Y_1, Y_2) = \rho(Y_1, Y_2) = \frac{\mathrm{cov}(Y_1, Y_2)}{\sqrt{\mathrm{var}(Y_1)\,\mathrm{var}(Y_2)}}.$$

The random variables are *uncorrelated* if $\rho = 0$. Independent random variables are by definition uncorrelated, but uncorrelated random variables need not be independent.

The above definitions all extend to the case of continuous random variables in the obvious way, i.e. by defining the density function to be any non-negative function over a range R of real valued pairs (y_1, y_2) whose integral over this range is equal to unity, and replacing summation everywhere by integration.

Finally, we can consider joint distributions of more than two variables, but the only important fact for later reference is that the random variables $Y_1, Y_2, \ldots, Y_n$ are said to be *mutually independent* if their joint density function is equal to the product of the n marginal density functions at all points of their range of definition.

If we have a sample of n values $y_1, y_2, \ldots, y_n$ from a population, then *sample* moments take the place of population moments. We will be much concerned with these sample statistics in the rest of this book, but for present purposes we note the two most common such moments:

(i) the sample mean $\bar{y} = \frac{1}{n}\sum_{i=1}^{n} y_i$,
(ii) the sample variance $s^2 = \frac{1}{n-1}\sum_{i=1}^{n}(y_i - \bar{y})^2$,

with sample standard deviation s being defined, as expected, to be the square root of the sample variance.

2.3 Some common distributions

We said in the previous section that to provide a probability model for a population we need to specify the probability density function $f(y)$, and that this is usually done by choosing an appropriate functional form for it. Then any observed data can be used to adjust parameter values of the function so that the model is as accurate as possible for the situation in hand. It is thus crucial to select an appropriate functional form in the first place. In principle, *any* function $f(y)$ satisfying the two conditions in section 2.1 can be chosen. However, in practice there will usually be certain features of the data and/or sampling process that make some functions more appropriate than others. In particular, there are about a dozen specific functions that supply models for populations in the overwhelming majority of practical situations, and these common distributions are now summarized. As always, it is convenient to group these distributions according as the random variable is discrete or continuous. In each case we give a brief sketch, indicating situations in which the distribution is applicable as a model, stating the density function and interpreting its parameters where necessary, giving the mean and variance, and concluding with a brief numerical example. (Readers who wish to test their manipulative skills might like to derive the means and variances from their definitions for each distribution.)

2.3.1 Discrete distributions

The discrete uniform distribution

This is almost a trivial model, as it simply puts the 'equally likely' interpretation of probability into the random variable context. Suppose that Y can take values $1, 2, \ldots, n$ and that each of these values is equally likely. Then

(i) $f(y) = \frac{1}{n}$ $(y = 1, 2, \ldots, n)$;
(ii) $\mu = (n + 1)/2$;
(iii) $\sigma^2 = (n^2 - 1)/12$.

The only parameter here is n, the upper limit of the range of values of y. This model may be useful for the national lottery, but is otherwise of limited direct practical applicability.

Example 2.2. If Y denotes the score on the upper face of a randomly thrown die, then Y has the distribution above with $n = 6$. Thus $\mu = 3.5$ and $\sigma^2 = 35/12 = 2.916$.

The Bernoulli distribution

As we saw in Chapter 1, many variables we deal with in practical applications are binary or dichotomous, i.e. the outcome can only be one of two types. A common way of categorizing the two outcomes is by scoring them 0 or 1, but a convenient way of thinking about them is as 'success' or 'failure'. A *Bernoulli trial* is a model for this simple situation, and the *Bernoulli distribution* gives the probabilities of outcomes from a single Bernoulli trial. If p is the probability of obtaining a success (and hence $1 - p$ the probability of obtaining a failure) then

   (i) $f(y) = p^y(1 - p)^{1-y}$ $(y = 0, 1)$;

   (ii) $\mu = p$;

   (ii) $\sigma^2 = p(1 - p)$.

Example 2.3. In the investigation of distances travelled from home to Exeter by first-year students, suppose that the researcher is interested in the proportion of such students who have to travel more than 100 miles. If a student is chosen at random and asked whether he or she travels more than 100 miles, 'yes' can be taken as a 'success' and 'no' as a failure so that this constitutes a Bernoulli trial with p equal to the proportion of all first-year students who have to travel more than 100 miles.

Single Bernoulli trials are relatively uninteresting. However, often we have a sequence of (mutually) independent Bernoulli trials in which the probability of success remains constant and for which we compute some single summary measure. Such situations lead to the next few distributions.

The binomial distribution

Suppose we have a fixed number n of independent Bernoulli trials in each of which the probability of success is p. Then if Y denotes the total number of successes observed, Y has the *binomial distribution* with

   (i) $f(y) = \binom{n}{y}p^y(1 - p)^{n-y}$ $(y = 0, 1, \ldots, n)$;

   (ii) $\mu = np$;

   (iii) $\sigma^2 = np(1 - p)$,

where $\binom{n}{y}$ denotes the combinatorial coefficient $\frac{n!}{y!(n-y)!}$ and ! is the factorial symbol (i.e. $n! = n(n-1)(n-2)\cdots 3\cdot 2\cdot 1$). This distribution, with parameters n and p, is extremely useful as a model for practical situations. Typically n will equal the size of a sample, and often interest focusses on the proportion of members that possess some attribute. Then we have n independent Bernoulli trials, each having constant probability p of success (where p is the proportion of individuals in the population that possess the attribute), and Y/n is the proportion of sample members that possess the attribute.

Example 2.4. Returning to Example 2.3, suppose that in fact 40% of all students have to travel more than 100 miles. Then if a sample of 20 students is interviewed, the number Y in the sample who have to travel more than 100 miles will have a binomial distribution with parameters $n = 20$ and $p = 0.4$. Thus, for example, the probability that exactly 5 sample members have to travel more than 100 miles is $\Pr(Y = 5) = f(5) = \binom{20}{5}(0.4)^5(0.6)^{15} = 0.0746$, while the probability that at least half the sample members have to travel more than 100 miles is $f(10) + f(11) + \cdots + f(20) = 0.2447$. The mean and variance of sample members that have to travel more than 100 miles are $20 \times 0.4 = 8$ and $20 \times 0.4 \times 0.6 = 4.8$ respectively.

The geometric distribution

In a sequence of Bernoulli trials each with probability of success p, let Y now denote the number of trials before the first success. Then Y has the *geometric distribution* with

(i) $f(y) = p(1 - p)^y$ $(y = 0, 1, \ldots)$;

(ii) $\mu = (1 - p)/p$;

(iii) $\sigma^2 = (1 - p)/p^2$.

Example 2.5. Continuing on from Example 2.4, suppose that students are interviewed sequentially. Then the probability that four students are interviewed before we encounter one who has to travel more than 100 miles is $\Pr(Y = 4) = f(4) = (0.4)(0.6)^4 = 0.0518$. The expected value of Y is $0.6 \div 0.4 = 1.5$, so with such a frequency of students travelling more than 100 miles, we would expect either the second or third one interviewed to have to do so.

The negative binomial distribution

Continuing the above theme, let Y now be the number of trials before the rth success in a sequence of independent Bernoulli trials each having probability p of success. Then Y has the *negative binomial distribution* with parameters r and p, for which

(i) $f(y) = \binom{y}{r-1} p^r (1 - p)^{y+1-r}$ $(y = r - 1, r, r + 1, \ldots)$;

(ii) $\mu = r(1 - p)/p$;

(iii) $\sigma^2 = r(1 - p)/p^2$.

The distinction between this distribution and the binomial distribution should be carefully noted. In the binomial distribution we perform a fixed number of trials and count the number of successes obtained, while in the negative binomial we fix the number of successes and count the number of trials needed to obtain them. It is also evident by comparison with the previous distribution that the negative binomial can be thought of as a sum of r independent geometric distributions, and it can easily be verified that all the formulae for the negative binomial reduce to those of the geometric when $r = 1$. (Note also that slightly different formulations of both geometric and negative binomial distributions result if the trial at which the defining success occurs is included in the count).

Example 2.6. Continuing with the set-up of Example 2.5, the probability that we interview four students before we encounter the *second* one who has to travel more than 100 miles is $f(4) = \binom{4}{1}(0.4)^2(0.6)^3 = 4 \times 0.16 \times 0.216 = 0.138$.

The Poisson distribution

In many practical situations, we effectively need to consider a very large number of Bernoulli trials n, for each of which the probability of success p is very small but the product np is neither excessively large nor vanishingly small. For example, the number of misprints on a page of typescript, the number of mutant cells observed on a plate under a microscope, or the number of albino mice in a large number of litters have these features. By taking the binomial distribution and applying mathematical limiting theory in the appropriate way, we arrive at the *Poisson distribution* for which

(i) $f(y) = \frac{\lambda^y e^{-\lambda}}{y!}$ $(y = 0, 1, 2, \ldots)$;
(ii) $\mu = \lambda$;
(iii) $\sigma^2 = \lambda$.

Example 2.7. A switchboard receives, on average, 20 calls an hour. What is the probability that between 30 and 36 calls come in the next two hours?

Since the average per hour is 20 calls, then the average per two-hour period is 40 calls. This is a good situation to model by a Poisson distribution, since we can break the two hours down into very many short instances in which a call might start, with very low probability that one will start in any given single instant. Moreover, the mean of a Poisson distribution is equal to its parameter [item (iii) above]. Hence if Y denotes the number of calls received in the next two hours, we can assume Y to have a Poisson distribution with parameter $\lambda = 40$. Thus the required probability is $f(30) + f(31) + \cdots + f(36) = \frac{(40)^{30}e^{-40}}{30!} + \frac{(40)^{31}e^{-40}}{31!} + \cdots + \frac{(40)^{36}e^{-40}}{36!} = 0.2531$.

Comment

In general, each of the above distributions has a defining feature which suggests which one might be suitable as a model in any particular given situation. Thus if we are concerned with proportions of individuals who possess certain attributes then by thinking of these attributes as 'successes' and the individual sample measurements as Bernoulli trials, we are led to the binomial distribution as a suitable model. Similarly, if we are concerned with counting 'rare events', then a natural model is the Poisson distribution. However, in some situations we may not have such obvious guidance and we will need to choose a model more empirically. In this case, the relationship between the mean and variance of the distribution is a useful indicator. Thus we see immediately from above that the mean and variance are equal for the Poisson distribution, so the Poisson is a good empirical model for discrete response variables which show approximate equality of sample mean and variance. By contrast, since both p and $1 - p$ lie between 0 and 1, the mean is greater than the variance for a binomial variable and less than the variance for a geometric variable. Inspection of the mean and variance of any available sample, therefore, provides a pointer to a reasonable model in any particular situation.

The multinomial distribution

Bernoulli variables have just two possible outcomes, 'success' or 'failure', and many practical situations can be modelled by calling on such variables. However, it often happens that observed variables have more than two categories. We have already seen that such cases can be handled by defining a number of dummy binary variables, so in principle we could model the data by a collection of Bernoulli variables. On the other hand, we frequently only need to consider a *summary* of the data, given by the number of observations in the sample that fall into each category. In this case using a set of Bernoulli variables is too cumbersome, and we would be better served by a generalization of the binomial distribution to the case of more than two categories.

Such a generalization is provided by the *multinomial* distribution. Suppose that the response can be in one of k categories, and that the probabilities of finding an observation in each of these categories are $p_1, p_2, \ldots, p_k$. Then if we have a sample of size n, the probability of observing n_1 individuals in category 1, n_2 individuals in category 2, and so on, is given by

$$f(n_1, n_2, \ldots, n_k) = \frac{n!}{n_1! n_2! \cdots n_k!} p_1^{n_1} p_2^{n_2} \cdots p_k^{n_k}.$$

However, since the categories are mutually exclusive and exhaustive we must have $p_1 + p_2 + \cdots + p_k = 1$, so that we can replace p_k by $1 - p_1 - p_2 - \cdots - p_{k-1}$. Also, we must have $n_1 + n_2 + \cdots + n_k = n$, so that $n_k = n - n_1 - n_2 - \cdots - n_{k-1}$. Substituting for both p_k and n_k in the above probability, we therefore have

$$f(n_1, n_2, \ldots, n_{k-1}, n) = C p_1^{n_1} p_2^{n_2} \cdots p_{k-1}^{n_{k-1}} (1 - p_1 - \cdots - p_{k-1})^{n - n_1 - \cdots - n_{k-1}}$$

where $p_i \geq 0$ for all i, $p_1 + \cdots + p_{k-1} \leq 1$, and

$$C = \frac{n!}{n_1! n_2! \cdots n_{k-1}! (n - n_1 - n_2 - \cdots - n_{k-1})!}.$$

This is the most parsimonious parameterization of the multinomial distribution, showing that it has $k - 1$ free parameters if there are k categories possible for the response. There are various properties associated with this distribution, but the only one we will need in the sequel is that the expected number in the jth category is given by np_j for $j = 1, \ldots, k$.

2.3.2 Continuous distributions

The uniform distribution

The continuous analogue of the discrete uniform distribution is defined in such a way that the probability of a uniform random variable falling in a given interval is directly proportional to the width of that interval. If Y is such a variable, defined on the range $a \leq y \leq b$, then

(i) $f(y) = \begin{cases} \frac{1}{b-a} & \text{if } a \leq y \leq b \\ 0 & \text{otherwise;} \end{cases}$

(ii) $\mu = (a + b)/2$;

(iii) $\sigma^2 = (b - a)^2/12$.

This distribution is very useful as a model for random phenomena associated with continuous measurements such as time, length, area etc. It also finds much use in simulating observations from such processes.

Example 2.8. A cloth-cutting machine can be set to cut a specific length of cloth. It cuts to the nearest centimetre, ensuring that the length is never less than the setting but leaving a random residue of up to 1 centimetre length. What is the probability that this residue lies between 0.35 and 0.72 centimetres?

From the description, it is reasonable to model the length in centimetres of the residue by a uniform distribution on the interval $(0, 1)$. Hence $a = 0$ and $b = 1$ in the above definition, so $f(y) = 1$ over the range $0 \leq y \leq 1$ and zero otherwise. Thus, from the defining feature of continuous variables (see Section 2.1), $\Pr(0.35 \leq Y \leq 0.72) = \int_{0.35}^{0.72} 1 \, dy = 0.72 - 0.35 = 0.37$.

The normal distribution

This is the most commonly used distribution in statistics. It is very conveniently parameterized directly in terms of its mean and its variance, and these two moments completely determine the whole distribution. If Y has a normal distribution with mean μ and variance σ^2, then it is defined over the whole real line and its probability density function is

$$f(y) = \frac{1}{\sigma\sqrt{2\pi}} e^{-\frac{1}{2}\left(\frac{y-\mu}{\sigma}\right)^2} \qquad (-\infty < y < \infty).$$

The notation $Y \sim N(\mu, \sigma^2)$ is often used to denote a variable with this distribution. It is easy to show that Y has skewness γ equal to zero, and kurtosis κ equal to 3.

To *standardize* a normal variable Y, we subtract its mean and divide the result by its standard deviation. This converts Y into the variable $Z = \frac{Y-\mu}{\sigma}$, which has mean zero and variance 1. The probability density function of Z is thus

$$f(z) = \frac{1}{\sqrt{2\pi}} e^{-\frac{1}{2}z^2} \qquad (-\infty < z < \infty),$$

and since this function is so important in statistics it is given the special symbol $\phi(z)$. It has a symmetric 'bell-shaped' curve which is effectively zero outside the range $-3 \leq z \leq +3$. Likewise, the distribution function derived from it has the special symbol Φ:

$$\Phi(z) = \int_{-\infty}^{z} \frac{1}{\sqrt{2\pi}} e^{-\frac{1}{2}u^2} \, du,$$

and this function has been extensively tabulated to enable probabilities of normal variables to be calculated accurately. (Note, however, that tables usually only give probabilities for positive values of z. Values for negative z must therefore be deduced from considerations of symmetry.)

Example 2.9. Suppose that the distance Y travelled by first-year students from home to Exeter can be assumed to have a normal distribution with mean 80 (miles) and standard deviation 20. What is the probability that a student chosen at random lives between 70 and 100 miles of Exeter?

We require $\Pr(70 \leq Y \leq 100)$. Standardizing, we see that this is equivalent to

$$\Pr\left(\frac{70 - 80}{20} \leq \frac{Y - 80}{20} \leq \frac{100 - 80}{20}\right),$$

i.e.

$$\Pr(-0.5 \leq z \leq 1.0) = \Pr(z \leq 1.0) - \Pr(z \leq -0.5).$$

From tables of the normal distribution we find this probability to be $0.8413 - 0.3085 = 0.5328$.

The exponential distribution

This random variable is defined on all positive values and has a single parameter. It is often used to model the time between events occuring at random, or the time to failure of simple electronic devices, and its density function is maximum at zero with exponential decay thereafter. If Y follows the exponential distribution then

(i) $f(y) = \lambda e^{-\lambda y}$ $(0 \leq y < \infty)$;
(ii) $\mu = 1/\lambda$;
(iii) $\sigma^2 = 1/\lambda^2$.

Example 2.10. The life in hours of a light bulb has an exponential distribution with mean 1000 hours. What is the probability that its time to failure lies between 500 and 1500 hours?

Since the mean is 1000, then $1/\lambda = 1000$ so that $\lambda = 0.001$. Thus the probability density function is $f(y) = 0.001e^{-0.001y}$ and the required probability is

$$\Pr(500 \leq Y \leq 1500) = \int_{500}^{1500} 0.001e^{-0.001y} \, dy = e^{-0.5} - e^{-1.5} = 0.3834.$$

The gamma and beta distributions

These are two distributions which provide extremely flexible modelling possibilities. Both have two parameters, varying which can generate a wide selection of shapes of density curves; the gamma distribution is defined on all positive real values, the beta distribution on the interval $(0, 1)$.

First, the gamma distribution. This has probability density function

$$f(y) = \frac{1}{\Gamma(b)} a^b y^{b-1} e^{-ay} \qquad (0 \leq y < \infty),$$

where $\Gamma(b)$ is the gamma function that is defined by $\Gamma(t) = \int_0^\infty x^{t-1} e^{-x} \, dx$ for $t > 0$ and has the property that $\Gamma(t + 1) = t\Gamma(t)$ [so that $\Gamma(n) = (n - 1)!$ if n is an integer].

The mean and variance of the gamma distribution are given by $\mu = b/a$ and $\sigma^2 = b/a^2$ respectively. For $b \leq 1$ the density function is J-shaped, otherwise it is unimodal and skewed to the right. The degree of skewness decreases with increasing b. The parameter a does not influence shape but just scales the distribution. When b is an integer, this distribution results from the sum of b independent exponential distributions each having density ae^{-ay}.

The beta distribution has density

$$f(y) = \frac{1}{B(a, b)} y^{a-1} (1 - y)^{b-1} \qquad (0 \leq y \leq 1),$$

where $B(a, b)$ is the beta function that is defined by $B(a, b) = \int_0^1 x^{a-1} (1 - x)^{b-1} \, dx$ for $a > 0$, $b > 0$ and has the property that $B(a, b) = B(b, a) = \frac{\Gamma(a)\Gamma(b)}{\Gamma(a+b)}$.

The mean and variance of the beta distribution are given by $\mu = a/(a + b)$ and $\sigma^2 = ab/[(a + b + 1)(a + b)^2]$ respectively. Both parameters influence the shape of the density. If $a > 1$ and $b > 1$ the distribution is unimodal, being symmetric if $a = b$, skewed to the right if $a > b$ and skewed to the left if $a < b$. The amount of skewness increases with the discrepancy between a and b. If $a \leq 1$ and $b > 1$ the density is J-shaped, whereas if $a > 1$ and $b \leq 1$ it is the opposite, i.e. the density increases monotonically with y. If both a and b are less than one it is U-shaped, while if both a and b equal one then it becomes the uniform distribution on $(0, 1)$.

Distributions derived from the normal distribution

There are three distributions which can be derived from the normal distribution and which occur very frequently in all branches of statistics. Their definitions are fairly straightforward and their cumulative distribution functions are extensively tabulated, but other aspects such as formulae for probability density functions and moments are of relatively minor importance. We therefore simply give their definitions here, and introduce their uses as necessary in succeeding sections.

(i) If $Y_1, Y_2, \ldots, Y_k$ are mutually independent $N(0, 1)$ variables, then

$$W = \sum_{i=1}^{k} Y_i^2$$

is said to have the chi-squared distribution on k degrees of freedom. We write $W \sim \chi_k^2$.

(ii) If Y has an $N(0, 1)$ distribution, W has a χ_k^2 distribution, and Y is independent of W, then

$$U = \frac{Y}{\sqrt{W/k}}$$

is said to have the (Student's) t-distribution on k degrees of freedom. We write $U \sim t_k$.

(iii) If $W_1 \sim \chi_k^2$, $W_2 \sim \chi_m^2$ and W_1, W_2 are independent, then

$$V = \frac{W_1}{k} \div \frac{W_2}{m}$$

is said to have the F-distribution on k and m degrees of freedom. We write $V \sim F_{k,m}$.

2.4 Sampling distributions

The probability distributions of the previous section provide models for the behaviour of response variables in a variety of situations. If we were to take a sample of values from one of these distributions, then the histogram of values observed would resemble the probability density function of the distribution, and the resemblance would increase as the sample size increased. (In fact, the probability density function would exactly match the outline of the histogram if every member of the population for which the distribution is the model had been sampled.)

Now when we take a random sample from a population whose probability density function is $f(y)$, then the randomness property ensures that the separate values observed are *mutually independent* while the fact that they come from the same population means that each is a possible value governed by the density $f(y)$. Taking a random of size n from $f(y)$ is therefore equivalent to observing the value of each of n separate random variables $Y_1, Y_2, \ldots, Y_n$ which are *independently and identically distributed (iid)* with probability density function $f(y)$. Much of statistical theory is accordingly concerned with a consideration of sets of iid variables. By way of shorthand we can write Y to denote the vector $(Y_1, Y_2, \ldots, Y_n)$ containing all n such variables, and y to denote the vector of values $(y_1, y_2, \ldots, y_n)$ observed on these variables.

However, we are rarely in practice interested in the set of separate values $y_1, y_2, \ldots, y_n$. More usually we first compute some summary value (such as the mean, variance, proportion of successes etc.) and then focus on this summary value in all our subsequent investigations. A *statistic* $T(Y)$ is any quantity that can be evaluated uniquely given the value of $Y = (Y_1, Y_2, \ldots, Y_n)$. Thus all the previously mentioned summary values are statistics in this sense, since mean $T_1 = \overline{Y} = \frac{1}{n} \sum_{i=1}^{n} Y_i$, variance $T_2 = S^2 = \frac{1}{n-1} \sum_{i=1}^{n} (Y_i - \overline{Y})^2$, proportion of successes $T_3 = P = \frac{1}{n} \sum_{i=1}^{n} Y_i$ if $Y_i = 0, 1$ for each i and $Y_i = 1$ denotes a success. Note that we have written all these statistics with upper case letters and have retained upper case Y_i in their definition, to denote that they are values which *might* occur from any particular population. The corresponding lower case versions ($t_1 = \overline{y} = \frac{1}{n} \sum_{i=1}^{n} y_i$, $t_2 = s^2 = \frac{1}{n-1} \sum_{i=1}^{n} (y_i - \overline{y})^2$, $t_3 = p = \frac{1}{n} \sum_{i=1}^{n} y_i$) would denote the *actual values* of these statistics obtained for the actual sample values $y_1, y_2, \ldots, y_n$. (One can think of T as being a *recipe* for a calculation, and t as the *outcome* of applying this recipe to a set of observations.)

This distinction is important, because a quantity such as $\overline{Y}$ or S^2 is itself a random variable. Since I cannot predict in advance what the actual values of the Y_i will be, then clearly I cannot predict in advance what the values of either $\overline{Y}$ or S^2 [or, indeed, of any other statistic $T(Y)$] will be. Moreover, if I take a sample of size n from a particular population and calculate the value of a given statistic T today, then repeat the same procedure from the same population tomorrow, and again the next day and so on for, say, a month, then the values $t_1, t_2, \ldots, t_{30}$ that I obtain will themselves come from some particular distribution of values. This distribution is known as the *sampling distribution* of the statistic T. It represents the population of possible values of T and the probabilities associated with them, and it plays a central role in all aspects of statistical inference.

The importance of the sampling distribution lies in the fact that it provides some measure of precision of any value of the statistic actually observed in practice. Any particular study will just yield one value of any statistic, and in isolation it is difficult to know how this value is to be interpreted. For example, suppose that a sample of 25 first-year students at Exeter University yields a mean distance of travel from home to Exeter of 85 miles, while a similar sample of first-year students at Birmingham University produces a mean distance of travel from home to Birmingham of 79 miles. We would be tempted to conclude that, on average, Exeter students live further away from university than do Birmingham students. But is this a valid conclusion? Are the two samples 'typical' or not? What sort of values might we get if we were to repeat the process, say, 20 times? Would the average of the averages still give the same conclusion? To answer these questions, we need to have some idea of what might happen in future samples of size 25 from the same populations, i.e. what features are exhibited by the sampling distributions of the mean distance in samples of size 25 from the two cities.

Deriving the sampling distribution of a statistic depends on the statistic itself, and in some cases is extremely difficult. Fortunately, if we assume that we are sampling from a normal population then some simple results can be obtained. Moreover, in the case of the most common statistics such as totals and means, there is a remarkable general result known as the central limit theorem which enables sampling distributions to be approximated very well in large samples irrespective of the nature of the population from which the original sample was taken.

First let us assume that we have taken a random sample of size n from a normal population that has mean μ and variance σ^2. This is equivalent to assuming that $Y_1, Y_2, \ldots, Y_n$ are iid $N(\mu, \sigma^2)$ variables. Then if $a_1, a_2, \ldots, a_n$ are constants it can be shown that $T = \sum_{i=1}^{n} a_i Y_i \sim N(\mu \sum_{i=1}^{n} a_i, \sigma^2 \sum_{i=1}^{n} a_i^2)$. This general result enables sampling distributions of common statistics to be derived. For example, if T_1 denotes the sum of values in the sample, then $T_1 = \sum_{i=1}^{n} Y_i$ so that $a_i = 1$ for all i and it follows that the sampling distribution of the sum is normal with mean $n\mu$ and variance $n\sigma^2$. Similarly, if T_2 denotes the mean of the sample values then $T_2 = \frac{1}{n} \sum_{i=1}^{n} Y_i$ so that $a_i = \frac{1}{n}$ for all i and it follows that the sampling distribution of the mean is normal with mean μ and variance σ^2/n. The standard deviation of the sampling distribution of a statistic T is known as the *standard error* of that statistic.

Example 2.11. As in Example 2.9, let Y denote the distance travelled by first-year students from home to Exeter and assume that it has a normal distribution with mean 80 and standard deviation 20. Now suppose that $\overline{Y}$ denotes the mean of a random sample of 25 such students. Then the above result tells us that the sampling distribution of $\overline{Y}$ is also normal, with mean 80 and variance $20^2/25 = 16$. The standard error of $\overline{Y}$ is thus $\sqrt{16} = 4$.

The probability that the sample mean lies between 70 and 100 is therefore

$$\Pr(70 \leq \overline{Y} \leq 100) = \Pr\left(\frac{70 - 80}{4} \leq \frac{\overline{Y} - 80}{4} \leq \frac{100 - 80}{4}\right),$$

i.e.

$$\Pr(-2.5 \leq z \leq 5.0) = \Pr(z \leq 5.0) - \Pr(z < -2.5),$$

and from tables this is $1.0 - 0.00621 = 0.99379$. (Contrast this with the probability of 0.5328, derived in Example 2.9, that a *single observation* in the sample will lie between 70 and 100.)

Next relax the assumption of sampling from a normal distribution, and merely suppose that $Y_1, Y_2, \ldots, Y_n$ are iid with mean μ, variance σ^2 but unspecified distribution. Then the central limit theorem tells us that the sampling distribution of the mean, $\overline{Y} = (Y_1 + Y_2 + \cdots + Y_n)/n$, converges in probability to $N(\mu, \sigma^2/n)$ as $n \to \infty$. The practical implication of this result is that if n is sufficiently large, then the sampling distributions of the sample total and mean are *approximately* those of T_1 and T_2 above *whatever the original population from which the sample is taken*. How large the sample has to be before the approximation is a good one depends on the population from which the sample is taken: if this population is symmetric and continuous then good results can be obtained with n as low as 5 or 6, but if the population is heavily skewed or otherwise problematic then we may need to have n around 30 before the approximation becomes acceptable.

Example 2.12. Consider the life in hours of light bulbs introduced in Example 2.10. From the properties of the exponential distribution we see that the variance is the square of the mean. A single light bulb therefore has a lifetime which is exponentially distributed with mean 1000 and variance 10^6. Using the central limit theorem, the average lifetime of a sample of 100 bulbs is thus approximately normally distributed, with mean 1000 and variance $10^6 \div 100 = 10^4$, i.e. a standard error of 100. The probability that the average lifetime of bulbs in the sample lies between 800 and 1200 hours is thus approximately 0.95.

The behaviour over repeated samples of a sample mean $\overline{Y}$ is important and worth a comment. If the mean and variance of the original population are μ and σ^2 respectively, then we see that the sampling distribution of $\overline{Y}$ is centred at the same mean μ but has less spread about the centre (standard error $\sigma/\sqrt{n}$ as opposed to σ). Moreover, this spread progressively reduces as the size of the sample from which $\overline{Y}$ is computed increases. Thus, for large samples, we would expect means of samples usually to be close to μ, even though some individual observations may be far from μ – large deviations from μ in one direction have a good chance of being balanced by large deviations in the other to give an average near the middle. Also, as the sample size increases, the shape of the sampling distribution approaches that of the normal distribution regardless of the shape of the parent population, and this is very important when it comes to inference (see below). The final point worth stressing is that a sample proportion can always be viewed as an average of 0/1 binary variables ('absence' or 'presence' of the attribute in question), so that sampling distributions of proportions also exhibit normality in large samples.

2.5 Inference

As already stressed earlier, we are usually concerned with making statements about a population but are limited to collecting information from just a sample (and often a small sample) of individuals from this population. We have seen that parametric models exist for most populations that we might encounter in practice, and generally speaking the statements we wish to make will concern the parameters of these models. For example, a parasitologist might have measured under a microscope the lengths of samples of parasites taken from two different species, and may wish to draw comparisons between the species in terms of their lengths. If the lengths of parasites can be assumed to be normally distributed in each species, with means μ_1 and μ_2 but a common variance σ^2 in the two species respectively, then a comparison between the two species is effected by comparing the two parameters μ_1 and μ_2.

The process of making statements about population parameters given only information from samples is known as *parametric statistical inference*. The sort of statements that we wish to make about parameters will often fall into one of a small number of categories, so methods of statistical inference have traditionally been divided into these categories. They are:

(i) Point estimation.
(What is the 'best guess' at the true value of a population parameter?)
(ii) Interval estimation.
(Can we find an interval of values within which we are 'fairly sure' that the true value of the population parameter lies?)
(iii) Hypothesis testing.
(Do the sample values 'support' or 'refute' a particular theory that we might hold?)

Thus the parasitologist may want to estimate the difference between the mean lengths of the two species (i.e. the value of $\mu_1 - \mu_2$). Alternatively he may wish to determine a range of values within which this difference is almost certain to lie. Finally, he may have a theory that individuals of the first species are on average longer than those of the second species (i.e. $\mu_1 > \mu_2$) and may wish to see if this theory is upheld by the sample data. Note that each of the questions above contains a word or phrase in inverted commas. This is because these phrases are ones that are understandable in everyday usage but need to be defined more precisely for use in a scientific context. We will therefore consider briefly each of these types of inference, outlining the methods involved and clarifying the interpretation of the phrases highlighted above.

2.5.1 Point estimation

Point estimation is the aspect of statistical inference in which we wish to make an 'informed guess' at the value of a population parameter, θ say. Since we have to use only sample data, and since we have to produce a single number as the value of θ, it is natural to use some statistic T to deliver this value. Such a statistic is called an *estimator* of θ, whilst the value t of the statistic observed for a particular sample is called an *estimate* of θ. However, in some situations it may not be clear how to find such a statistic while in others there may be several available possibilities and we need some criterion for choosing between them. Also, once we have chosen a statistic, we will want to know something about its behaviour. There are thus two prime questions that arise:

- How do we find an estimator?
- How good is our resulting estimate likely to be?

The former concerns the *method* of estimation, while the latter requires study of *properties* of estimators. We consider each of these questions in turn.

Methods of estimation

Sometimes, an 'obvious' estimator will suggest itself. For example, a population mean is most naturally estimated by a sample mean, and a population proportion by the corresponding sample proportion. Thus the mean distance travelled in miles between their homes and Exeter by all first-year students at Exeter University would obviously be estimated by taking a random sample of first-year students at Exeter and calculating the mean distance travelled by the sample members. Similarly, the proportion unemployed in a sample of males over the age of 18 taken from the electoral roll of a particular town would be the obvious estimator of the proportion of over-18 males that were unemployed in this town.

Often, however, the situation is more complicated and an estimator is not at all obvious. For example, suppose that n randomly chosen light bulbs have lifetimes (in hours) $y_1, y_2, \ldots, y_n$. I buy a new light bulb, and want to estimate the probability that it will last longer than h hours. A simple approach would say that probability is the same as population proportion, so we want to estimate the proportion of individuals in the population that last more than h hours. A suitable estimator would thus be the proportion in the sample that last longer than h hours. However, this will be in general a very crude estimate, particularly if n is small or h is large (or both) – there is only a restricted set of available estimates, of the form i/n for $i = 0, 1, 2, \ldots, n$, with an estimated probability of zero as the most likely outcome. To make better progress we need to model the situation. We saw earlier that lifetimes of light bulbs are well modelled by an exponential distribution which has the probability density

$$f(y) = \lambda e^{-\lambda y} \qquad (0 \leq y < \infty).$$

Adopting this as a model we find $\Pr(y > h) = \int_h^\infty \lambda e^{-\lambda y} \, dy = e^{-\lambda h}$, so this is the quantity to be estimated from the sample. This boils down to estimating λ from the sample and substituting the estimate into this expression. So how can we find an estimator of λ?

There are three popular methods of estimation that provide guidance in such cases: the methods of *moments*, *maximum likelihood*, and *least squares*. Broadly speaking, the method of moments is the simplest of the three and works well in straightforward cases but doesn't always produce estimators with good properties, maximum likelihood is a good general-purpose method which produces estimators that have excellent large-sample properties but sometimes poor small-sample properties, while least squares is best in the context of linear models and is equivalent to maximum likelihood when the data come from normal distributions. We will describe each method in turn and illustrate it with examples.

The method of moments

If the population parameter to be estimated is the population mean, then we have said above that the 'obvious' estimator is the sample mean. The method of moments simply takes this logical argument one step further, by choosing the estimate of any general unknown parameter to be that value which makes the population mean equal to the sample mean. For any particular probability model, the required estimator can be derived very easily. First the population mean is found in terms of the unknown parameter. Next this expression is set equal to the sample mean. Finally, the resulting equation is solved for the unknown parameter; this is the method of moments estimator of the parameter.

Example 2.13. Consider the light bulbs example outlined above. This is equivalent to saying that $Y_1, Y_2, \ldots, Y_n$ are iid variables, each with probability density function $f(y) = \lambda e^{-\lambda y}$ ($0 \leq y < \infty$). The mean of this distribution, from Section 2.3.2 above, is $1/\lambda$. Thus if $\overline{Y}$ is the mean of the Y_i then we set $\overline{Y} = 1/\lambda$ and solve for λ, to obtain $\tilde{\lambda} = 1/\overline{Y}$ as the method of moments estimator of λ. (Consequently we would estimate the probability that a future light bulb has lifetime greater than h hours by $e^{-h/\overline{Y}}$.)

Several points arise from this example. The first is to notice that an *estimator* (or estimate) of a parameter is indicated by placing some sort of mark over the parameter in question. We have used a tilde in the example above, and will denote method of moments estimators by this symbol throughout the book.

The second point is that the population mean is the first moment $E(Y)$ of a distribution, and $\overline{Y} = \frac{1}{n} \sum_{i=1}^{n} Y_i$ is its sample equivalent. The name of the method thus derives from our equating the sample and population moments to get the estimator. This principle extends if there are several unknown parameters in the model, but of course we need more than one equation to solve in this case. If there are two population parameters then we need to use the first *two* moments to obtain the estimator, viz. set $E(Y)$ equal to $\frac{1}{n} \sum_{i=1}^{n} Y_i$ and $E(Y^2)$ equal to $\frac{1}{n} \sum_{i=1}^{n} Y_i^2$ and solve the resulting pair of simultaneous equations. If there are three parameters we use the first *three* moments in this way, and so on.

The method of maximum likelihood

Let $Y_1, Y_2, \ldots, Y_n$ be a random sample from any probability density function $f(y; \theta)$ that depends on an unknown parameter θ. The *likelihood* of the sample is the joint probability density $f(y_1, y_2, \ldots, y_n; \theta)$ of the sample, *treated as a function of* θ. Writing the likelihood therefore as $L(\theta)$, and noting that in a random sample the individuals are independent of each other (so the joint density is the product of the individual densities), we have

$$L(\theta) = \prod_{i=1}^{n} f(y_i; \theta). \tag{2.15}$$

The function $\hat{\theta} = g(y_1, y_2, \ldots, y_n)$ that maximizes $L(\theta)$ with respect to θ is the *maximum likelihood estimator* of θ, and its actual value for a given sample is the *maximum likelihood estimate* (mle) of θ for that sample. Very roughly, the mle can be interpreted as the value of θ that ascribes the highest possible probability to the sample that was actually obtained, which is an intuitive justification for its use. Note also that the value maximizing $L(\theta)$ maximizes the (natural) logarithm of $L(\theta)$ as well, and the latter maximization is often easier to effect in practice. In 'regular' situations (i.e. continuous parameter, twice differentiable log-likelihood function, range of Y independent of θ) the mle can be found using standard calculus. If we write $l(\theta) = \log L(\theta)$ then the mle is that root of the equation $\frac{\partial l}{\partial \theta} = 0$ at which $\frac{\partial^2 l}{\partial \theta^2}$ is negative. If more than one such root exists, then we take the root at which the value of $L(\theta)$ is largest.

Example 2.14. Consider again the situation in Example 2.13: $Y_1, Y_2, \ldots, Y_n$ are iid variables, each with probability density function $f(y) = \lambda e^{-\lambda y}$ ($0 \le y < \infty$). Then

$$L(\lambda) = \prod_{i=1}^{n} f(y_i; \lambda),$$

$$= \prod_{i=1}^{n} \lambda e^{-\lambda y_i},$$

$$= (\lambda)^n \exp\left\{-\lambda \sum_{i=1}^{n} y_i\right\}.$$

Thus

$$l(\theta) = \log L(\theta) = n \log \lambda - \lambda \sum_{i=1}^{n} y_i,$$

so that

$$\frac{\partial l}{\partial \theta} = \frac{n}{\lambda} - \sum_{i=1}^{n} y_i$$

and setting the right-hand side equal to zero and solving for λ yields

$$\hat{\lambda} = \frac{n}{\sum_{i=1}^{n} y_i} = \frac{1}{\bar{y}}.$$

Moreover,

$$\frac{\partial^2 l}{\partial \theta^2} = -\frac{n}{\lambda^2},$$

which must be negative at $\lambda = \hat{\lambda}$ (since λ^2 must be positive for all $\lambda > 0$). Thus the maximum likelihood estimator of λ is given by $\frac{1}{\bar{Y}}$ (upper case variable for estimator), and we see from Example 2.14 that maximum likelihood yields the same estimator as the method of moments in this situation.

We will use a hat rather than a tilde to denote mles. If we have more than one parameter then the mles of the set of parameters are the values that jointly maximize the (log-)likelihood function. In regular cases they are found by setting to zero the partial derivatives of the log-likelihood with respect to each of the unknown parameters, and solving the resulting set of simultaneous equations.

Example 2.15. Suppose that $y_1, y_2, \ldots, y_n$ is a random sample from a normal distribution with mean μ and variance σ^2, two parameters we would like to estimate. Then

$$L(\mu, \sigma^2) = \prod_{i=1}^{n} \left(\frac{1}{\sigma\sqrt{2\pi}} \right) e^{-\frac{1}{2\sigma^2}(y_i - \mu)^2}$$

so that

$$
\begin{aligned}
l(\mu, \sigma^2) &= \log L(\mu, \sigma^2) \\
&= \log\left(\frac{1}{\sigma\sqrt{2\pi}} \right)^n - \frac{1}{2\sigma^2} \sum_{i=1}^{n} (y_i - \mu)^2 \\
&= -n \log \sigma - \frac{n}{2} \log(2\pi) - \frac{1}{2\sigma^2} \sum_{i=1}^{n} (y_i - \mu)^2.
\end{aligned}
$$

Hence

$$
\begin{aligned}
\frac{\partial l}{\partial \mu} &= \frac{1}{\sigma^2} \sum_{i=1}^{n} (y_i - \mu) \\
\frac{\partial l}{\partial \sigma} &= -\frac{n}{\sigma} + \frac{1}{\sigma^3} \sum_{i=1}^{n} (y_i - \mu)^2.
\end{aligned}
$$

Setting these equations to zero and solving them for σ and μ yields the mles $\hat{\sigma}$ and $\hat{\mu}$. Thus once we do set them to zero, we must change σ and μ to $\hat{\sigma}$ and $\hat{\mu}$ respectively.

Setting the first equation to zero (and multiplying through by $\hat{\sigma}^2$ which cannot be zero) yields $\sum_{i=1}^{n} (y_i - \hat{\mu}) = 0$, which gives $\hat{\mu} = \frac{1}{n} \sum y_i = \bar{y}$. Setting the second equation to zero (and multiplying through by $\hat{\sigma}^3$) yields $\hat{\sigma}^2 = \frac{1}{n} \sum (y_i - \hat{\mu})^2 = \frac{1}{n} \sum (y_i - \bar{y})^2$. These are the mles of μ and σ^2 respectively. Note the divisor n in $\hat{\sigma}^2$, as opposed to the divisor $n - 1$ in the usual definition of sample variance. We will comment on this point below.

The method of least squares

Let us for a moment reconsider two aspects of the situation in Example 2.15. First, if $Y_1, Y_2, \ldots, Y_n$ are iid $N(\mu, \sigma^2)$ variables, then we can equivalently write $Y_i = \mu + \epsilon_i$ for $i = 1, 2, \ldots, n$ where the ϵ_i are iid $N(0, \sigma^2)$ variables. In this formulation we can think of ϵ_i as the random variable which measures the *departure* of Y_i from the population mean μ. Second, we see that the log-likelihood has the form

$$l(\mu, \sigma^2) = -n \log \sigma - \frac{n}{2} \log(2\pi) - \frac{1}{2\sigma^2} \sum_{i=1}^{n} (y_i - \mu)^2,$$

so if σ^2 is simply regarded as an irrelevant constant then maximizing the log-likelihood with respect to μ is the same as minimizing $\sum_{i=1}^{n}(y_i - \mu)^2$ with respect to μ. In other words, the mle of μ is the value which minimizes the sum of squared departures between the sample values and the parameter.

The method of least squares generalizes this idea to functions of parameters and distributions other than the normal. If $Y_1, Y_2, \ldots, Y_n$ are such that $Y_i = g(\beta_1, \beta_2, \ldots, \beta_k) + \epsilon_i$ for $i = 1, 2, \ldots, n$, where $g(\beta_1, \beta_2, \ldots, \beta_k)$ is a constant function of k parameters β_i and the ϵ_i are iid random variables each with zero mean and a common variance σ^2, then the *least squares estimators* (lse) of the parameters are the values which minimize $V = \sum_{i=1}^{n}[Y_i - g(\beta_1, \beta_2, \ldots, \beta_k)]^2$. Standard calculus can again be employed to obtain these values. Evidently, least squares estimators coincide with maximum likelihood ones if the distribution of the ϵ_i is normal, but not necessarily otherwise. The method of least squares is at the heart of the techniques in the next few chapters, so we defer examples of its use till then.

Properties of estimators

Whenever we produce an estimate of an unknown parameter θ, it is natural to ask 'how accurate is it?' More generally, we might enquire what properties a 'good' estimator should have. Being able to compare their properties will enable us to choose between competing estimators.

Example 2.16. Suppose that $y_1, y_2, \ldots, y_n$ is a random sample from the uniform distribution on the interval $(0, \theta)$, i.e. the distribution with probability density function

$$f(y) = \frac{1}{\theta} \qquad 0 \le y \le \theta,$$

and we wish to estimate θ.

The population mean here is $E(Y) = \int_0^\theta y \frac{1}{\theta} \, dy$, which is easily shown to be $\theta/2$. Thus if the sample mean is denoted by $\bar{y}$, then the method of moments estimate is given by $\tilde{\theta} = 2\bar{y}$.

The likelihood of the sample, on the other hand, is given by

$$L = \frac{1}{\theta^n} \qquad (0 \le y_1, y_2, \ldots, y_n \le \theta).$$

This is maximized when θ takes its smallest possible value. But since, by definition, θ must be greater than or equal to any of the sample y_i, it cannot be smaller than the largest sample value. Standard notation denotes the ith largest (ranked) observation by $y_{(i)}$, so that the largest sample value is denoted $y_{(n)}$. Thus the maximum likelihood estimate is $\hat{\theta} = y_{(n)}$.

Here we have a situation in which the methods of moments and maximum likelihood give us *different* estimates. Which one should we choose?

Since any estimator can be viewed as a 'recipe' for providing estimates, one way of judging the worth of an estimator $\hat{\theta}$ is by looking at the set of all estimates it might produce in relation to the true value θ. To do this we need to consider the *sampling distribution* of $\hat{\theta}$ over repeated samples of size n from the same population. Several summary features of this distribution provide relevant properties of $\hat{\theta}$.

Bias

An estimator is said to be *unbiased* if $E(\hat{\theta}) = \theta$, i.e. if the mean of its sampling distribution is θ, otherwise it is *biased* with bias $b(\theta) = E(\hat{\theta}) - \theta$. An unbiased estimator is therefore one which gives the correct value 'on average' over repeated sampling, which is clearly a useful property for it to have.

Example 2.17. The method of moments estimator of θ for a random sample from the uniform distribution on $(0, \theta)$ was found in Example 2.16 to be $\tilde{\theta} = 2\bar{Y} = \frac{2}{n} \sum_{i=1}^{n} Y_i$. Thus

$$E(\tilde{\theta}) = \frac{2}{n} \sum_{i=1}^{n} E(Y_i) = \frac{2}{n} n \left(\frac{\theta}{2} \right) = \theta$$

(since each Y_i has expected value equal to $\theta/2$, the mean of the uniform distribution). Thus the method of moments estimator is an *unbiased* estimator of θ.

Finding the expected value of the maximum likelihood estimator requires knowledge of the sampling distribution of the largest member $Y_{(n)}$ of a random sample of size n from a uniform distribution, which goes beyond the level of theory in this book. However, it can be shown that $E(\hat{\theta}) = \frac{n\theta}{n+1}$. Hence the maximum likelihood estimator is biased, with bias equal to $\frac{n\theta}{n+1} - \theta = -\frac{\theta}{n+1}$. (Note that this bias tends to zero as n tends to infinity, a point we return to when considering properties of mles below.)

In a similar way we can establish that the maximum likelihood estimator $\frac{1}{n} \sum_{i=1}^{n} (y_i - \bar{y})^2$ of σ^2 in a random sample of size n from an $N(\mu, \sigma^2)$ distribution is biased, but the usual sample variance $s^2 = \frac{1}{n-1} \sum_{i=1}^{n} (y_i - \bar{y})^2$ is an unbiased estimator of σ^2. This explains the preference for the latter estimator in practical applications.

Mean square error

An unbiased estimator merely ensures that the average of a (large) series of repeated estimates is correct, but says nothing about individual estimates. As in practice we usually only take *one* sample, it would be nice to be sure that this *particular* estimate was close to θ. The mean square error (MSE) of an estimator $\hat{\theta}$ is defined to be $E(\hat{\theta} - \theta)^2$, i.e. the average squared distance of $\hat{\theta}$ from the true value θ over repeated sampling, so a 'good' estimator is one with a *small* MSE. If we have a choice of several estimators, the best one will be the one with smallest MSE. Moreover, it can be shown that MSE is identically equal to squared bias plus variance. Hence the MSE of an unbiased estimator is equal to its variance, so the usual characterization of a 'good' estimator is as an unbiased estimator with small variance. However, there is also a famous result known as the *Cramer–Rao Lower-Bound Theorem* which establishes that the variance of any unbiased estimator of a parameter θ cannot be less than $I^2(\theta)$, where

$$I^2(\theta) = \frac{1}{-E\left(\frac{\partial^2 l}{\partial \theta^2}\right)}, \tag{2.16}$$

and, as usual, l denotes the log-likelihood of the sample providing the estimator. Thus no unbiased estimator can do 'better' than one whose variance is equal to $I^2(\theta)$, and hence such an estimator is said to be *fully efficient*.

Example 2.18. We saw in Example 2.15 that the sample mean $\overline{Y}$ is the maximum likelihood estimator of μ in a random sample from an $N(\mu, \sigma^2)$ distribution, and in Section 2.4 it was shown that the sampling distribution of $\overline{Y}$ is normal with mean μ and variance σ^2/n. Thus $\overline{Y}$ is an unbiased estimator of μ. Also, from Example 2.15 we have

$$\frac{\partial l}{\partial \mu} = \frac{1}{\sigma^2} \sum_{i=1}^{n} (y_i - \mu),$$

so that

$$\frac{\partial^2 l}{\partial \mu^2} = -\frac{1}{\sigma^2} \sum_{i=1}^{n} (1) = -\frac{n}{\sigma^2}.$$

Since this expression just involves the constants n and σ^2, $E\left(\frac{\partial^2 l}{\partial \mu^2}\right) = -\frac{n}{\sigma^2}$ and hence $I^2(\mu) = \frac{\sigma^2}{n}$. Thus $\overline{Y}$ is a fully efficient estimator of μ.

Consistency

The preceding properties all involve the sampling distribution of an estimator $\hat{\theta}$ obtained from a sample of size n. This sampling distribution reflects values that would be obtained for $\hat{\theta}$ if repeated samples *of size n* were taken from the same population. To check that we have a sensible estimator, however, we need also to satisfy ourselves that $\hat{\theta}$ is increasingly likely to yield the right answer θ as the sample size gets bigger and, in the limiting case, is *certain* to yield θ if the whole population is sampled. This latter condition is satisfied if the MSE is zero, so we say that $\hat{\theta}$ is a *consistent* estimator of θ if its MSE $\to 0$ as $n \to \infty$. If $\hat{\theta}$ is unbiased then it is also consistent if its variance $\to 0$ as $n \to \infty$.

Example 2.18 (cont.). It is evident from Example 2.18 that $\text{var}(\overline{Y}) = \sigma^2/n \rightarrow$ 0 as $n \rightarrow \infty$, so $\overline{Y}$ is a consistent estimator of μ.

Maximum likelihood estimators

There are two very nice properties of maximum likelihood estimators that make these estimators theoretically very attractive.

The first is known as the *invariance* property, which says that if $\hat{\theta}$ is the mle of θ and $h(\theta)$ is any 1–1 continuous function of θ then $h(\hat{\theta})$ is the mle of $h(\theta)$. This enables us to deduce mles for most functions of parameters we encounter in practice. For instance, we saw in Example 2.15 that $\hat{\sigma}^2 = \frac{1}{n} \sum (y_i - \overline{y})^2$ is the mle of the variance σ^2 of a normal population. The invariance property thus allows us to deduce that $\sqrt{\frac{1}{n} \sum (y_i - \overline{y})^2}$ is the mle of the population standard deviation σ. (Note that only the positive square root is allowed in the definition of standard deviation, so the transformation considered here is 1–1.)

The second property is that the sampling distribution of the mle $\hat{\theta}$ of a parameter θ converges in probability to the normal distribution with mean θ and variance $I^2(\theta)$ as the sample size $n \rightarrow \infty$. Thus every mle is asymptotically (i.e. for large n) unbiased and fully efficient. Moreover, providing n is not too small we can say that the sampling distribution of $\hat{\theta}$ is approximately $N(\theta, I^2(\theta))$, and the approximation improves as n increases.

2.5.2 Interval estimation

While the idea of point estimation is a very natural one, and obtaining a single 'best guess' of an unknown parameter value is a desirable aim in many practical situations, there is one serious problem with the whole procedure. All our criteria for judging the worth of an estimator are essentially based on long-run repeated sampling performances. Thus we acknowledge that, even though our chosen estimator may be the 'best' one possible in terms of these criteria, our estimate of the unknown parameter will *change from sample to sample*. This in turn implies that in any *single* sample, our estimate will *almost certainly be wrong*. But generally we *only* have a single sample in practice, so the pessimists will question the worth of the whole exercise.

Although such a view is no doubt overly pessimistic, nevertheless it may be preferable to sacrifice some precision if the benefit is greater confidence in the derived estimate. Thus, instead of finding a single value (which we are almost certain is *not* the true value of the parameter), perhaps it would be better to find a *range* of values within which we are almost certain that the true parameter value *does* lie. Such a range of values is known as a *confidence interval* for the unknown parameter. The *width* of this interval will in general depend on two factors: the *level of confidence* that we wish to ascribe to its containing the true parameter value, and the sampling variability of the statistics used to construct the interval. Clearly, the wider the interval, the more likely it is to contain the true parameter value. Equally, the more sampling variability there is in the point estimate of the parameter, the less precision attaches to any single estimate and so the wider must be any interval to ensure a given level of confidence that the true parameter has been included.

The general method for constructing a confidence interval has the following three stages:

(i) Identify a *pivotal function* from any suitable point estimator of the parameter in question. (A pivotal function is a function of the unknown parameter, sample statistics and constants whose distribution is known and for which probabilities can be found from tables. A necessary condition for the latter is that this distribution does not depend on any unknown parameter.)

(ii) Select a probability value, typically one of 0.99, 0.95, 0.90, and find critical values from tables such that the probability of the pivotal function lying between the critical values equals the chosen probability value.

(iii) Manipulate the probability statement implicit in (ii) so that the unknown parameter is in the centre of the inequality and only sample statistics or constants are at its extremes. The end points of the confidence interval are defined by these extremes, and the confidence level is given by the chosen probability value expressed as a percentage.

To clarify this general procedure, we now present as examples the derivation of some commonly needed confidence intervals.

Example 2.19. Suppose that $y_1, y_2, \ldots, y_n$ is a random sample from a normal distribution with mean μ and variance σ^2, where σ^2 is a known constant, and we want to find a confidence interval for μ. Consider each of the above stages in turn.

(i) From earlier sections (cf Example 2.18) we know that the sample mean $\overline{Y}$ is the maximum likelihood estimator of μ in a random sample from an $N(\mu, \sigma^2)$ distribution, and that the sampling distribution of $\overline{Y}$ is $N(\mu, \frac{\sigma^2}{n})$. Thus

$$Z = \frac{\overline{Y} - \mu}{\sigma/\sqrt{n}} \sim N(0, 1).$$

Now the quantity Z is a function of the unknown parameter (μ), sample statistics ($\overline{Y}$) and constants (n, σ^2), while its distribution $N(0, 1)$ is known and tabulated. Thus Z is a pivotal function according to the definition above.

(ii) Next we need the confidence level, so let us select the value 0.95, say. From normal tables we find the critical values $+1.96$ and -1.96 that satisfy $\Pr(-1.96 \leq Z \leq +1.96) = 0.95$ when $Z \sim N(0, 1)$.

(iii) Substituting the pivotal quantity from (i) for Z in the probability statement in (ii) we obtain

$$\Pr\left(-1.96 \leq \frac{\overline{Y} - \mu}{\sigma/\sqrt{n}} \leq +1.96\right).$$

Then multiplying throughout the inequality by $\sigma/\sqrt{n}$, subtracting $\overline{Y}$ from each side and multiplying throughout by -1, we obtain

$$\Pr\left(\overline{Y} - 1.96\frac{\sigma}{\sqrt{n}} \leq \mu \leq \overline{Y} + 1.96\frac{\sigma}{\sqrt{n}}\right) = 0.95.$$

We have now satisfied the conditions required of the probability statement (viz. only the unknown parameter μ in the centre, and only sample statistics ($\overline{Y}$) or constants ($1.96, n, \sigma$) at the extremes), so a 95% confidence interval for μ is given by the interval

$$\left(\overline{Y} - 1.96\frac{\sigma}{\sqrt{n}}, \overline{Y} + 1.96\frac{\sigma}{\sqrt{n}}\right).$$

Note in the above that whereas probabilities are generally quoted as proportions (0.95), confidence levels are usually given as percentages (95%). Also, the only difference on changing the confidence level is in the critical value. For 90% confidence we would have the value 1.645 instead of 1.96 (shorter interval so less confidence that it contains μ), while for 99% confidence we would have 2.5758 (longer interval so greater confidence that it contains μ). We can subsume all these values into one general formula by saying that

$$\left(\overline{Y} - z_\alpha\frac{\sigma}{\sqrt{n}}, \overline{Y} + z_\alpha\frac{\sigma}{\sqrt{n}}\right)$$

is a $100(1 - \alpha)$% confidence interval for μ, where z_α is the value from normal tables such that $\Pr(-z_\alpha \leq Z \leq +z_\alpha) = 1 - \alpha$ when Z is an $N(0, 1)$ variable.

Example 2.20. Suppose again that $y_1, y_2, \ldots, y_n$ is a random sample from a normal distribution with mean μ and variance σ^2 and we want to find a confidence interval for μ, but this time σ^2 is unknown.

If σ^2 is unknown, then it would seem reasonable to try the same process as above but replacing σ^2 wherever it occurs by its estimate, the sample variance s^2. Thus instead of Z, we should consider

$$T = \frac{\overline{Y} - \mu}{s/\sqrt{n}}.$$

Now the quantity T is again a function of the unknown parameter for which the confidence interval is required (μ), sample statistics ($s, \overline{Y}$) and constants (n). Moreover, basic statistical theory establishes that its sampling distribution is the t distribution on $n - 1$ degrees of freedom, which is also tabulated. Thus T is another pivotal function, appropriate for the present problem.

Now let $t_{n-1,\alpha}$ be the critical value from tables such that $\Pr(-t_{n-1,\alpha} \leq T \leq +t_{n-1,\alpha}) = 1 - \alpha$ when T has a t distribution on $n - 1$ degrees of freedom. Then substituting the pivotal quantity above for T we obtain

$$\Pr\left(-t_{n-1,\alpha} \leq \frac{\overline{Y} - \mu}{s/\sqrt{n}} \leq t_{n-1,\alpha}\right),$$

and carrying out analogous manipulations to those in Example 2.19 we obtain

$$\Pr\left(\overline{Y} - t_{n-1,\alpha}\frac{s}{\sqrt{n}} \leq \mu \leq \overline{Y} + t_{n-1,\alpha}\frac{s}{\sqrt{n}}\right) = 1 - \alpha$$

so that

$$\left(\overline{Y} - t_{n-1,\alpha}\frac{s}{\sqrt{n}}, \overline{Y} + t_{n-1,\alpha}\frac{s}{\sqrt{n}}\right)$$

is a $100(1 - \alpha)\%$ confidence interval for μ.

Example 2.21. Suppose again that $y_1, y_2, \ldots, y_n$ is a random sample from a normal distribution with mean μ and variance σ^2, but now we want to find a confidence interval for σ^2 when μ is unknown.

Another standard result of statistical theory tells us that if s^2 is the sample variance for a random sample from a normal distribution, then the sampling distribution of $W = \frac{(n-1)s^2}{\sigma^2}$ is chi-squared on $n - 1$ degrees of freedom. Thus W is again a pivotal quantity, so we can use the same sort of argument as before to derive a confidence interval for σ^2. However, there is just one important difference in this case. The standard normal and t distributions used in Examples 2.19 and 2.20 are both symmetric about zero, so only one critical value had to be found from tables and plus/minus this value was used in the confidence interval construction. The chi-squared distribution, by contrast, is not symmetric and is only defined on positive values of the variable. Thus we need to find *two* critical values from tables, c_1 and c_2 say, such that $\Pr(c_1 \leq W \leq c_2) = 1 - \alpha$. However, many different pairs satisfying this inequality can be found. For a unique choice, we therefore look for the values that cut off an equal proportion $(\alpha/2)$ in each tail of the distribution. If $\chi^2_{n-1,\alpha}$ denotes the value *exceeded* by a proportion α of the chi-squared distribution on $n - 1$ degrees of freedom, then the required choices are $c_1 = \chi^2_{n-1,1-\alpha/2}$ and $c_2 = \chi^2_{n-1,\alpha/2}$.
Then

$$\Pr\left(c_1 \leq \frac{(n-1)s^2}{\sigma^2} \leq c_2\right) = 1 - \alpha,$$

so that

$$\Pr\left(\frac{1}{c_2} \leq \frac{\sigma^2}{(n-1)s^2} \leq \frac{1}{c_1}\right) = 1 - \alpha$$

and hence

$$\Pr\left(\frac{(n-1)s^2}{c_2} \le \sigma^2 \le \frac{(n-1)s^2}{c_1}\right) = 1 - \alpha.$$

Thus $\left(\frac{[n-1]s^2}{c_2}, \frac{[n-1]s^2}{c_1}\right)$ is a $100(1-\alpha)\%$ confidence interval for σ^2.

Interpretation of confidence intervals

All intervals derived in the manner set out above have end points which are functions of sample statistics. Since every sample statistic has a sampling distribution, and its value will change from sample to sample, then any confidence interval will vary from sample to sample. If all the distributional assumptions underlying the construction of the interval are met, however, then this construction ensures that a proportion $100(1-\alpha)\%$ of all intervals will enclose the true value of the parameter they are estimating. This is the sense in which we are $100(1-\alpha)\%$ confident that any single interval, constructed for the one sample that is available in a practical situation, encloses the true value of the unknown parameter.

2.5.3 Hypothesis testing

A common situation is where an investigator has a theory about the phenomenon under study, and wishes to see whether this theory is substantiated by data that have been collected. If a statistical model can be formulated for the data and the theory can be expressed as values of the model parameters, then the amount of support for the theory that is provided by the data can be assessed by means of a *hypothesis test*. In practice, we can rarely achieve absolute evidence, but must adopt a comparative approach against some 'standard' or 'neutral' position. The latter is generally termed the *null hypothesis*, denoted H_0, while the theory to be tested is termed the *alternative hypothesis* and is usually denoted either H_1 or H_a. (We shall generally use H_a). Implicitly, the null hypothesis is what we are prepared to 'go along with' until we obtain convincing evidence in favour of the alternative.

Assuming that we have data available, and can formulate a suitable statistical model (which usually just means specifying the type of probability distribution from which the data have come), then to conduct a hypothesis test we need to complete the following steps.

(i) Specify the null and alternative hypotheses.

(ii) Choose a *test statistic* T, say, which is such that

- T behaves *differently* under the null and alternative hypotheses, and
- the sampling distribution of T is fully specified when H_0 is true.

(iii) Assess the weight of evidence in favour of H_a over H_0.

We first amplify each of these stages, introducing technical terms as necessary, and then illustrate the procedure with some examples.

To specify the null and alternative hypotheses, we need to attach values to model parameters. A hypothesis is said to be *simple* if a single value is attached to each unknown parameter, and *composite* otherwise. The latter can arise either because a *range* of values is attached to a parameter, or because one or more parameters are left unspecified. Parameters left unspecified are called *nuisance* parameters. The most common situation is where there is just a single parameter θ, say, and the null hypothesis specifies that it is equal to a single known value θ_0, say. Then a composite alternative hypothesis can be either *one-sided* ($\theta > \theta_0$ or $\theta < \theta_0$) or *two-sided* ($\theta \neq \theta_0$).

In simple situations, choice of test statistic can be guided either by estimators of the parameter in question or by pivotal functions involving them. We illustrate this in the examples below. In more complicated situations, the likelihood ratio principle generally yields a suitable test statistic (see below).

To assess the weight of evidence in favour of H_a over H_0, let us suppose that the sample data produce a value t of the test statistic T. Then we assume that H_0 is true and calculate the probability p of observing a value of T that is either as extreme as t or more extreme than t *in the direction of departure of H_a from H_0*. For example, if θ is the mean of a normal distribution and H_0 specifies that $\theta = 0$ while H_a specifies that $\theta > 0$, then values of T 'more extreme than t in the direction of H_a' would be values such that $T > t$. On the other hand, if H_a were $\theta \neq 0$, then 'more extreme' would be *either $T > |t|$ or $T < -|t|$*. The value of p is known as the *significance level* of the test. Roughly speaking it tells us how likely we are to observe a value at least as favourable to H_a as the one we have observed, if H_0 is true. The *smaller* this value is, the less likely is H_0 to be true and hence the *greater* is the evidence in favour of H_a. If p is *sufficiently small* then one is justified in *rejecting H_0* in favour of H_a. The values 0.05 and 0.01 have become traditional in this context; if $p < 0.05$ (0.01) we say that the test is 'significant at the 5% (1%) level'.

Note that the p value gives the probability of observing a value as extreme as the one we have, *when the null hypothesis is true*. Thus if we decide always to reject H_0 if, say, $p < 0.05$ (i.e. if the test is significant at the 5% level), then we will reject *incorrectly* on 5% of occasions. Rejecting H_0 when it is true is known as a type 1 error. To reduce the chance of making this error, we need to reduce the significance level at which we decide to reject H_0 (say to 1%). However, making the conditions for rejection more stringent in this way also reduces the chances of *not* rejecting H_0 when it *should* be rejected, i.e. when H_a is true. This latter error is known as a type 2 error, and the fundamental problem with hypothesis testing is that, in general, acting to reduce the probability α of type 1 error automatically increases the probability β of type 2 error. The compromise reached in practice is to fix α at a small value, hence the traditional 0.05 or 0.01, and if there is a choice of tests to pick the one with smallest β. The probability α is often called the *size* of the test, while $1 - \beta$ (i.e. the probability of rejecting H_0

when H_0 is indeed false) is known as the *power* of the test.

One point that needs to be appreciated when testing hypotheses is that the p value (and, in general, the conclusion from the test) depends on the size of any samples used in the procedure. As sample sizes increase it becomes easier to reject a null hypothesis. The reason for this is that if the true value of the tested parameter varies from the null hypothesis by even a minimal amount, we will be able to detect this difference if our samples are large enough. Thus it should always be remembered that a *significant* result may not necessarily be a *practically important* one!

Example 2.22. Suppose I believe that a certain coin is biased towards 'heads' when spun. I spin the coin 12 times, and observe 10 'heads'. Is this sufficient evidence to substantiate my belief?

Here, a suitable statistical model would be given by assuming the 12 spins of the coin to be 12 independent Bernoulli trials, each having probability π of 'success' (i.e. of landing 'heads'). Then, if Y denotes the number of 'heads' observed in the 12 trials, from Section 2.3.1 we see that Y has a binomial distribution with probability function

$$f(y) = \binom{12}{y} \pi^y (1 - \pi)^{12-y} \qquad (y = 0, 1, \ldots, 12).$$

The null hypothesis here, i.e. the hypothesis that would seem reasonable until evidence to the contrary is obtained, is clearly that the coin is unbiased, i.e. that $\pi = 0.5$. The alternative hypothesis is the one into which we place our belief, so here it is that $\pi > 0.5$. Thus we wish to test

$$H_0 : \pi = 0.5 \quad \text{vs} \quad H_a : \pi > 0.5.$$

Given that these hypotheses pertain to observation of 'heads', Y itself is an obvious test statistic to adopt.

The last step of the procedure is to calculate the probability of obtaining a value as extreme as, or more extreme than, the one we have observed, in the case where H_0 is true. We have observed $Y = 10$, and values more favourable to H_a than this are $Y = 11$ and $Y = 12$, while if H_0 is true then $\pi = 0.5$. Thus, the significance level of our test is

$$p = \binom{12}{10} 0.5^{10} 0.5^2 + \binom{12}{11} 0.5^{11} 0.5^1 + \binom{12}{12} 0.5^{12} 0.5^0$$
$$= 0.0161 + 0.0029 + 0.0002 = 0.0192.$$

We would consider this to be strong evidence in favour of our belief. If we take a formal rejection/not rejection view, then we would reject H_0 here at the 5% level of significance but not (quite) at the 1% level.

Note that if our belief was simply that the coin was biased, but without any feeling as to whether the bias was in favour of 'heads' or 'tails', then the alternative hypothesis would be the two-sided $\pi \neq 0.5$ instead of the one-sided $\pi > 0.5$. A result as extreme or more extreme than the one observed would thus now be $Y = 0, 1, 2, 10, 11, 12$ (as we might equally have observed 10, 11 or 12 'tails' as 'heads' and it is just chance that our particular sample has yielded 'heads'). The significance level would now be the probability of observing any of these values if $\pi = 0.5$, and since the binomial distribution in this case is symmetric then $\Pr(Y = 0$ or 1 or $2) = \Pr(Y = 10$ or 11 or $12) = 0.0192$, so that $p = 0.0384$. Since we use probabilities from both ends of the distribution, this is called a 'two-tailed' test in contrast to the 'one-tailed' test that was used for the one-sided $H_a : \pi > 0.5$.

Example 2.23. A sugar-packing machine produces packets whose weights are normally distributed with mean μ grams and standard deviation 2.5 grams. Regulations demand that μ be no larger than 1001. A sample of 20 packets is obtained, and the sample mean weight of the packets is $\bar{y} = 1002$ g. Is this significant evidence that the regulations are being breached?

Here the statistical model is already specified, namely $Y \sim N(\mu, 2.5^2)$ where Y is the weight of an individual packet of sugar. The largest tolerable value of μ is 1001, while anything greater than this contravenes regulations, so obvious choices of null and alternative hypotheses would be $H_0 : \mu = 1001$ and $H_a : \mu > 1001$ respectively.

A common way of selecting a test statistic is to look at the pivotal quantity for the confidence interval calculation in the same probability model, and to replace the unknown parameter by its null hypothesis value. From Example 2.19 we see that the pivotal quantity for the mean of a normal population whose variance is known is

$$Z = \frac{\bar{Y} - \mu}{\sigma/\sqrt{n}},$$

so replacing μ by its null hypothesis value 1001, σ by its known value 2.5, and n by the sample size 20, the test statistic here becomes

$$Z = \frac{\bar{Y} - 1001}{2.5/\sqrt{20}}.$$

Moreover, if H_0 is true then the sampling distribution of Z is $N(0, 1)$. Thus the probability of obtaining a value as extreme or more extreme than the observed $\bar{Y} = 1002$ is equal to the probability that a standard normal variable exceeds

$$Z = \frac{1002 - 1001}{2.5/\sqrt{20}} = 1.789,$$

and from tables of the normal distribution we see that this probability equals 0.037. Thus the significance level of the test is 0.037, which provides moderately strong evidence that the regulations are being breached (significant at the 5% but not at the 1% level).

Example 2.24. Children's heights (in inches) can be assumed to be normally distributed. A sample of 20 six-year-old girls gave sample mean $\overline{y} = 44.85$ and sample standard deviation $s = 3.39$. Do these data support the view that the mean height of all six-year-old girls is 43 inches?

Here we assume that if Y denotes a child's height then $Y \sim N(\mu, \sigma^2)$ and we wish to test $H_0 : \mu = 43$ versus $H_a : \mu \neq 43$ with σ unspecified (i.e. a nuisance parameter). Referring to the corresponding confidence interval situation, we find the pivotal quantity

$$T = \frac{\overline{Y} - \mu}{s/\sqrt{n}}$$

which has a t distribution on $n - 1$ degrees of freedom when μ has its true value. Thus replacing μ by its null hypothesis value of 43 yields our test statistic, and substituting the sample values $\overline{y} = 44.85$, $s = 3.39$, $n = 20$ gives the observed value of this statistic as $t = 2.44$ on 19 degrees of freedom.

Since the test is two-tailed (a two-sided alternative hypothesis), the significance value is $p = \Pr(T < -2.44 \text{ or } T > 2.44)$. Unfortunately, since the t distribution is different for every different number of degrees of freedom, tables of the distribution are not detailed enough for us to be able to find the p value exactly. The best that can be done is to find t_a values such that $p_a = \Pr(T < -t_a \text{ or } T > t_a)$ for specified values of p_a (such as 0.1, 0.05, 0.01, etc). We find from these tables that $t_a = 2.093$ for $p_a = 0.05$ and $t_a = 2.861$ for $p_a = 0.01$. Interpolating between these two values, we would estimate our significance level for $t = 2.44$ to be about 0.025. Alternatively, looking at the two t_a values, we would reject H_0 at the 5% significance level (since $2.44 > 2.093$, so that p must be less than 0.05) but we would not reject it at the 1% level (since $2.44 < 2.861$, so that p must be greater than 0.01).

Likelihood ratio test statistics

It is not always possible to deduce a sensible test statistic from either a point estimator of the parameter in question or a pivotal function involving it, especially when the situation under consideration is not one of the common, straightforward ones. Thus we need some systematic procedure that will generate a test statistic for us, whatever the situation or model under consideration. We have already seen in Section 2.5.1 that the likelihood can play a key role in obtaining estimators of unknown parameters, and we now show how the likelihood can also be employed to derive test statistics.

Let us suppose quite generally that θ denotes the set of unknown parameters, Ω denotes the space of all possible values of these parameters (i.e. the complete parameter space) and ω denotes the space of possible values of these parameters permitted by the null hypothesis (i.e. the restricted parameter space). To illustrate, in the situation of Example 2.23 we have $\theta = (\mu, \sigma^2)$, Ω is the set of all (μ, σ^2) pairs satisfying $-\infty < \mu < \infty$ and $0 \leq \sigma^2 < \infty$, and ω is the set of all (μ, σ^2) pairs in which μ is set at 43 but σ^2 can take any value between 0 and ∞.

To derive the *generalized likelihood ratio test statistic* in any situation we:

(i) find the supremum (maximum) of the likelihood when H_0 is true, i.e. $\sup_\omega L(\boldsymbol{\theta})$;

(ii) find the supremum (maximum) of the likelihood unconditionally, i.e. $\sup_\Omega L(\boldsymbol{\theta})$;

(iii) obtain the ratio of these two quantities,

$$\Lambda = \frac{\sup_\omega L(\boldsymbol{\theta})}{\sup_\Omega L(\boldsymbol{\theta})}$$

and reduce it down to the simplest statistic that is a monotonic function of Λ. This statistic is the required test statistic.

To illustrate the procedure, let us continue with the set-up in Example 2.23. The likelihood there is

$$L(\mu, \sigma^2) = \frac{1}{(\sigma\sqrt{2\pi})^n} \exp\left\{-\frac{1}{2\sigma^2} \sum_{i=1}^{n} (y_i - \mu)^2\right\}$$

with $n = 20$. Maximizing this likelihood unconditionally (i.e. over Ω) gives the usual maximum likelihood estimates $\hat{\mu} = \bar{y}$ and $\hat{\sigma}^2 = \frac{1}{n}\sum(y_i - \bar{y})^2$ (see Example 2.15), and when we substitute these values back in to the likelihood we obtain

$$\sup_\Omega L(\boldsymbol{\theta}) = \frac{1}{(\hat{\sigma}\sqrt{2\pi})^n} e^{-n/2}.$$

When the null hypothesis is true then $\mu = 43$ and the likelihood is a function of σ^2 only, viz

$$L(\sigma^2) = \frac{1}{(\sigma\sqrt{2\pi})^n} \exp\left\{-\frac{1}{2\sigma^2} \sum_{i=1}^{n} (y_i - 43)^2\right\}.$$

Maximizing this restricted likelihood with respect to σ^2 we find the maximum is at $\tilde{\sigma}^2 = \frac{1}{n}\sum(y_i - 43)^2$, and substituting this value back in to the restricted likelihood we obtain

$$\sup_\omega L(\boldsymbol{\theta}) = \frac{1}{(\tilde{\sigma}\sqrt{2\pi})^n} e^{-n/2}.$$

Hence the likelihood ratio test statistic is given by

$$\Lambda = \left(\frac{\hat{\sigma}^2}{\tilde{\sigma}^2}\right)^{n/2}$$

on cancelling all common factors.

Now by adding and subtracting $\bar{y}$ inside the square, we can write $\tilde{\sigma}^2 = \frac{1}{n}\sum[(y_i - \bar{y}) + (\bar{y} - 43)]^2$. On expanding the square in terms of the factors in round brackets the cross-product term vanishes (because $[\bar{y} - 43]$ is constant while $\sum_i [y_i - \bar{y}] = 0$), and we obtain $\tilde{\sigma}^2 = \hat{\sigma}^2 + (\bar{y} - 43)^2$. A small amount of manipulation reduces the above statistic into the monotonically equivalent form

$$n(\Lambda^{-2/n} - 1) = \frac{n(\bar{y} - 43)^2}{\hat{\sigma}^2},$$

and the right-hand side of this equation is just the square of the 'natural' statistic T given in Example 2.23.

In this example, therefore, the generalized likelihood ratio test statistic coincides with the usual test statistic, but this need not be the case for other situations. We will see later that the generalized likelihood ratio test plays an important role in statistical modelling.

2.6 Postscript

This concludes our brief summary of those elements of probability and statistics that are necessary for an understanding of statistical modelling, and we now go on to develop a range of modelling techniques. The next two chapters deal with those techniques that rest on assumptions of normality for the response variable, while the rest of the book widens the scope to include a selection of other distributions. Where necessary, we refer to the relevant sections of the present chapter in which particular results are quoted.

3
Normal response and quantitative explanatory variables: regression

3.1 Motivation

Consider the following example. Patients admitted to a clinic undergo a routine screening procedure, and various measurements are made on them in the laboratory. A group of 24 patients with hyperlipoproteinaemia, a metabolic disorder characterized by high levels of lipoproteins in the blood that can be associated with coronary heart disease, are observed to have the following plasma levels of total cholesterol (in mg/ml):

$$3.5, 1.9, 4.0, 2.6, 4.5, 3.0, 2.9, 3.8, 2.1, 3.8, 4.1, 3.0,$$
$$2.5, 4.6, 3.2, 4.2, 2.3, 4.0, 4.3, 3.9, 3.3, 3.2, 2.5, 3.3.$$

Suppose that the doctor wishes to predict the total cholesterol level of the next patient to be admitted to the clinic with hyperlipoproteinaemia. What would be the best way to do this?

Following the methods outlined in Chapter 2, it would seem reasonable to treat the response 'total cholesterol level' as a random variable Y (since it manifestly varies unpredictably from individual to individual), and the continuous nature of the measurements suggests the normal distribution as a plausible population model for Y. The mean and variance of this population are clearly unknown, so let us assume they are given by the parameters μ and σ^2 respectively. Thus the values given above may be treated as 24 independent observations $y_1, y_2, \ldots, y_{24}$ from an $N(\mu, \sigma^2)$ distribution. We may equivalently write any observation y_i as $\mu + \epsilon_i$ where $\epsilon_i \sim N(0, \sigma^2)$ independently of any other ϵ_j. In particular, the *next* observation to arise may be written as $y_{25} = \mu + \epsilon_{25}$. The 'best' estimate of ϵ_{25} is its expected value, i.e. zero, so the doctor's prediction of y_{25} will be μ. However, this quantity is unknown so must in turn be estimated from the 24 available observations, and all the methods of estimation discussed in Chapter 2 lead to the same estimate, namely $\hat{\mu} = \bar{y} = 3.354$.

What can we say about the accuracy of this prediction? First note that the predicted value is made up of two components, an estimate $(\bar{y})$ of μ plus an estimate (zero) of ϵ_{25}, and each of these estimates is subject to error. To assess these errors we note that $\hat{\mu}$ has variance $\sigma^2/24$ (by the results in Section 2.4), while ϵ_{25} has variance σ^2 (by assumption). Since the next (i.e. 25th) observation is independent of the preceding 24, $\hat{\mu}$ is independent of ϵ_{25}. The two earlier variances can thus be added, to give

$$\sigma^2\left(\frac{1}{24}+1\right) = 25\sigma^2/24$$

as the variance of the doctor's prediction. However, σ^2 is still unknown; its unbiased estimate (see the paragraph following Example 2.17) is given by

$$\hat{\sigma}^2 = \frac{1}{23}\sum_{i=1}^{24}(y_i - \bar{y})^2 = 0.6052,$$

so the doctor's prediction has estimated variance $(25 \times 0.6052)/24 = 0.6304$ and hence estimated standard error $\sqrt{0.6304} = 0.794$. Since we have used 23 degrees of freedom to estimate σ^2, reference to t-tables yields a 95% critical value of 2.069 and hence

$$3.354 \pm 2.0687 \times 0.794, \qquad \text{i.e. } (1.711, 4.997),$$

as the 95% confidence interval for the predicted value.

This is clearly a disappointingly wide confidence interval (which includes all 24 original values), so we conclude that the accuracy of prediction is rather low. However, in the absence of further information, it is the best that we can do. If we want to improve predictive accuracy, we need to collect further data that are relevant to the prediction we wish to make. Relevant data will in general come in the form of values on explanatory variables *that are related to the response variable, and that will somehow account for the observed variation in the response variable readings*. Incorporating such variables in the statistical model should lead to better prediction.

In the present example, it turns out that total cholesterol shows a tendency to increase with age, and so the ages of the 24 patients were also recorded. In the same order as the above cholesterol levels, these ages (in years) were:

$$46, 20, 52, 30, 57, 25, 28, 36, 22, 43, 57, 33,$$
$$22, 63, 40, 48, 28, 49, 52, 58, 29, 34, 24, 50.$$

Denoting 'age' by X with values $x_1, x_2, \ldots, x_{24}$, we see that each patient has associated with him/her a *pair* of values (x_i, y_i). Thus the data sample can be plotted in a *scattergram*, by associating the horizontal axis with X, the vertical axis with Y, and representing the ith patient by a point whose coordinates on these axes are x_i and y_i respectively. Such a plot is shown in Fig. 3.1.

There are two features worth noting about this plot and the associated data. The first is that there appears to be a roughly *linear* relationship between age

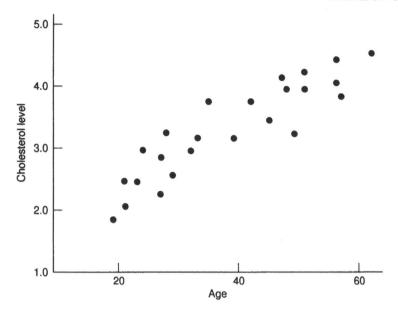

Fig. 3.1 *Plot of 24 patients' age against total cholesterol levels.*

and cholesterol level, for the range of values represented by these 24 patients. This means that if a straight line were fitted through the points, then most of the points would lie fairly close to the line. The general equation of a straight line is $y = \alpha + \beta x$, where α is the intercept (i.e. y value when $x = 0$) and β is the slope (i.e. increase in y for a unit increase in x). Thus if we determine a suitable α and β for this set of data, then given the age x_i of a patient the natural prediction of that patient's cholesterol level would be given by the corresponding point on the line, i.e. $\alpha + \beta x_i$.

The second feature is that, although there is considerable variability in the cholesterol values across the whole sample (between 1.9 and 4.6), there is much less variablity *within a particular age (or small range of ages)*. For example, there are two patients aged 22 and their cholesterol values are 2.1 and 2.5, two aged 52 with cholesterol values 4.0 and 4.3, and two aged 57 with cholesterol values 4.1 and 4.5. Alternatively, taking patients aged between 33 and 36 we find cholesterol values 3.0, 3.2 and 3.8. In each of these four situations the cholesterol values are closely bunched. Thus if we have found appropriate values of α and β above, then predicting the cholesterol level by $\alpha + \beta x_i$ should give a much more accurate result than that obtained without the help of age.

Putting these two features together points to the advantage of including patients' age in the predictive model. The simplest way of doing this is to assume that the mean cholesterol level is a linear function of age and that the individual patient cholesterol levels vary randomly about the mean appropriate to their age. This in turn requires only a small modification to the earlier model. There we assumed that a particular cholesterol observation y_i could be written as $y_i = \mu + \epsilon_i$ where $\epsilon_i \sim N(0, \sigma^2)$. Now all we need to modify is the population mean, and the simplest modification is to replace μ by $\mu_i = \alpha + \beta x_i$. In this way we maintain normality, independence and constant variance of observations, but allow each individual to come from a population whose mean depends on the individual's age in a linear fashion. This is known as the *simple linear regression model* (although the word 'linear' is often taken as understood).

However, while inclusion of one explanatory variable in this way might improve the predictive model, it usually happens that there is still some 'unexplained' variability left over so there is scope for increasing the accuracy of prediction yet further. In this case, extra explanatory variables need to be sought and included in the model. For example, 'average amount of alcohol consumed daily' might improve the predictions of cholesterol level even more if included with age in the model. In general we can envisage inclusion of p explanatory variables $X_1, X_2, \ldots, X_p$, and the simplest model that generalizes the previous ones is the model that retains all previous features but allows the means μ_i to be linear functions of *all* these explanatory variables. This is known as the *multiple linear regression model* (with the word 'linear' again being generally taken as understood).

Practical issues associated with these regression models include the fitting of the models to data (i.e. the estimation of the unknown parameters), the conducting of inferences about these parameters (i.e. constructing confidence intervals for them or carrying out hypothesis tests about them), the prediction of future values, the testing of model assumptions, the selection of 'best' models (i.e. selection of variables to include in models), the assessment of the worth of fitted models and the testing for lack of fit, the comparison of models between different groups or subgroups of data, and the question of whether more complex models than linear ones should be fitted. These issues are addressed in the present chapter.

3.2 Simple regression

We will use the cholesterol level and age data for illustration throughout this section, so it is convenient to collect the data in such a way that each patient's age (X) and cholesterol level (Y) are given side by side. This is done in Table 3.1.

3.2.1 The model

Given n pairs of data points $(x_1, y_1), (x_2, y_2), \ldots, (x_n, y_n)$, the discussion in the previous section has led to the model

$$y_i = \alpha + \beta x_i + \epsilon_i \qquad (i = 1, \ldots, n), \tag{3.1}$$

Table 3.1 *Age in years* (X) *and total cholesterol level in* mg/ml (Y) *for each of 24 patients with hyperlipoproteinaemia.*

X	Y	X	Y	X	Y	X	Y
46	3.5	20	1.9	52	4.0	30	2.6
57	4.5	25	3.0	28	2.9	36	3.8
22	2.1	43	3.8	57	4.1	33	3.0
22	2.5	63	4.6	40	3.2	48	4.2
28	2.3	49	4.0	52	4.3	58	3.9
29	3.3	34	3.2	24	2.5	50	3.3

where the ϵ_i are mutually independent random variables, each with zero mean and constant variance σ^2. Note that the requirement of constant variance implies that σ^2 does not depend on X; ways of checking this assumption are given in Section 3.2.6. Note also that normality of the ϵ_i is not specifically assumed at this stage; we comment on this aspect below. Finally, note that the form of the model implies that the only random variable is the response variable Y. The explanatory variable X is always taken to be *fixed*, i.e. measured without error.

The ϵ_i are commonly termed the *departures* of the y_i from their means $\alpha + \beta x_i$ for $i = 1, \ldots, n$, the reason for this name being illustrated in Fig. 3.2. (They are also often described as *errors*, but we shall not do so as this can be a misleading term.) Although the model given above is in a form that corresponds directly

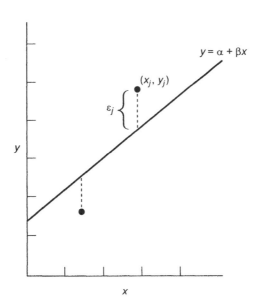

Fig. 3.2 *Illustration of the components of a simple regression model.*

with the equation of a straight line, for purposes of mathematical manipulation it is convenient to reparameterize it into the form

$$y_i = \beta_0 + \beta_1(x_i - \bar{x}) + \epsilon_i \qquad (i = 1, \ldots, n) \tag{3.2}$$

where $\bar{x}$ is, as usual, the sample mean of the x_i. It can be seen that the parameters of the two models are simply related by $\beta = \beta_1$ and $\alpha = \beta_0 - \beta_1\bar{x}$, so that either set can be obtained easily from the other. Computations are considerably simplified by adopting form (3.2), so this is the version used typically in fitting the model, but to summarize results it is best to quote the systematic part (cf Section 1.5) of the model in form (3.1) as this is the equation of the fitted line. Fitting the model is alternatively called *regressing Y on X*, and the fitted line is also known as the *regression line*.

3.2.2 Fitting the model

We look for optimal estimates $\hat{\beta}_0$ and $\hat{\beta}_1$ of the two unknown parameters in (3.2). Although it is possible to assume normality of the ϵ_i and thus use maximum likelihood to effect the estimation, it is preferable to make the model as widely applicable as possible by not assuming any specific distribution for the ϵ_i. In this case, least squares is a viable principle to apply for estimation (and, of course, it will give the same estimates as maximum likelihood if normality *is* assumed).

We thus look for estimates $\hat{\beta}_0$ and $\hat{\beta}_1$ that minimize the sum of squared departures,

$$S = \sum_{i=1}^{n} \epsilon_i^2 = \sum_{i=1}^{n} (y_i - \beta_0 - \beta_1[x_i - \bar{x}])^2.$$

Differentiating, we obtain

$$\frac{\partial S}{\partial \beta_0} = \sum 2(y_i - \beta_0 - \beta_1[x_i - \bar{x}])(-1)$$

$$\frac{\partial S}{\partial \beta_1} = \sum 2(y_i - \beta_0 - \beta_1[x_i - \bar{x}])(-[x_i - \bar{x}]).$$

Setting these partial derivatives to zero (for the minimum), and noting that $\sum_i [x_i - \bar{x}] = 0$, yields the *normal equations* for the minimizing values:

$$n\hat{\beta}_0 = \sum y_i$$

$$\sum (x_i - \bar{x})^2 \hat{\beta}_1 = \sum (x_i - \bar{x}) y_i.$$

But $\sum_i (x_i - \bar{x}) y_i = \sum_i (x_i - \bar{x})(y_i - \bar{y})$ since $\bar{y} \sum_i (x_i - \bar{x}) = 0$, so that if we adopt the shorthand notation

$$S_{xx} = \sum (x_i - \bar{x})^2, \quad S_{yy} = \sum (y_i - \bar{y})^2, \quad S_{xy} = \sum (x_i - \bar{x})(y_i - \bar{y})$$

then the normal equations can be written

$$n\hat{\beta}_0 = n\overline{y}$$
$$S_{xx}\hat{\beta}_1 = S_{xy},$$

and the parameter estimates follow directly from these equations as

$$\hat{\beta}_0 = \overline{y}, \quad \hat{\beta}_1 = S_{xy}/S_{xx}. \tag{3.3}$$

The simplicity of solution of these equations follows specifically from the use of $(x_i - \overline{x})$ in model (3.2). Differentiating S a second time with respect to each of the parameters yields

$$\frac{\partial^2 S}{\partial \beta_0^2} = 2n \quad \text{and} \quad \frac{\partial^2 S}{\partial \beta_1^2} = 2S_{xx}.$$

These second derivatives are everywhere positive, showing that (3.3) indeed provides the minimum of S.

Example 3.1. For the age and cholesterol level values in Table 3.1, we find

$$\sum x_i = 946, \quad \sum y_i = 80.5, \quad S_{xx} = 4139.833, \quad S_{yy} = 217.858,$$

so that $\hat{\beta}_0 = 80.5/24 = 3.354$ and $\hat{\beta}_1 = 217.858/4139.833 = 0.0526$. Thus $\hat{\alpha} = \hat{\beta}_0 - \hat{\beta}_1\overline{x} = 3.354 - 0.0526 \times 946/24 = 3.354 - 2.073 = 1.281$, so that the equation of the fitted regression line is $y = 1.281 + 0.0526x$.

It should be noted that in some circumstances the data are such as to force the regression line to go through the origin. This happens when the physical situation demands that $y_i = 0$ when $x_i = 0$ (e.g. when Y is the extension of a vertical spring and X is the weight of an object attached to it). In this case the model (3.1) might be replaced by

$$y_i = \beta x_i + \epsilon_i \quad (i = 1, \ldots, n). \tag{3.4}$$

This is often called the 'no-intercept' model, and direct application of least squares in this case yields the estimate $\hat{\beta} = (\sum_i x_i y_i)/(\sum_i x_i^2)$. However, care should be exercised in using this model. Sometimes it seems as though it should be appropriate (e.g. when Y is the height of a plant and X is the number of days since planting), but one can never be sure unless observations are available across a wide range of X values *including those near $X = 0$*. If such values are not available, then there is always a danger of non-linear behaviour near $X = 0$ and fitting the no-intercept model will be misleading.

3.2.3 Assessing the regression

Having fitted a regression line as above, the first thing we would like to assure ourselves is that using it will help to provide better predictions. In other words, is the relationship between Y and X beneficial in predicting Y? To answer this question, let us first introduce some useful quantities. Given *any* statistical model of the general form

$$y_i = \text{systematic component} + \epsilon_i \qquad (i = 1, \ldots, n),$$

the ith *residual* e_i is defined to be

$$e_i = y_i - \hat{y}_i,$$

where $\hat{y}_i$ is the value fitted (i.e. predicted) by the model at $X = x_i$ for $i = 1, \ldots, n$. The residuals e_i are thus the intuitive estimates of the departures ϵ_i, and if all the residuals are small then the model can be said to fit the data well. The sum of squared residuals $\sum_{i=1}^{n} e_i^2$ should therefore provide a good single summary measure of adequacy of the model, so we now consider the use and behaviour of this measure.

In the absence of information on X, we see from the discussion in Section 3.1 that the implicit model for the data is

$$y_i = \mu + \epsilon_i$$

for $i = 1, \ldots, n$, in which case the estimate of μ is the sample mean $\bar{y}$. Hence

$$\hat{y}_i = \bar{y}$$

for all i, so that

$$e_i = y_i - \bar{y}$$

and the sum of squared residuals is

$$E_0 = \sum_{i=1}^{n} (y_i - \bar{y})^2 = S_{yy}.$$

If we fit the simple regression model (3.2), then y_i is 'predicted' by the point on the regression line corresponding to $X = x_i$, so that

$$\hat{y}_i = \hat{\beta}_0 + \hat{\beta}_1 (x_i - \bar{x}) = \bar{y} + \frac{S_{xy}}{S_{xx}} (x_i - \bar{x})$$

on substituting from (3.3) for the parameter estimates. Thus

$$e_i = y_i - \hat{y}_i = (y_i - \bar{y} - \frac{S_{xy}}{S_{xx}} [x_i - \bar{x}]),$$

so that the sum of squared residuals is

$$E_1 = \sum_{i=1}^{n} \left([y_i - \bar{y}] - \frac{S_{xy}}{S_{xx}} [x_i - \bar{x}] \right)^2.$$

On squaring out the bracket and summing we find

$$E_1 = S_{yy} + \frac{S_{xy}^2}{S_{xx}^2} \times S_{xx} - 2 \times S_{xy} \times \frac{S_{xy}}{S_{xx}},$$

which reduces to

$$E_1 = S_{yy} - \frac{S_{xy}^2}{S_{xx}}.$$

(Clearly, this value E_1 is the same as the value of S in Section 3.2.2 when $\hat{\beta}_0$, $\hat{\beta}_1$ are substituted for β_0, β_1 respectively, so E_1 is the minimum possible value of S for the given data.)

From the above expressions it is evident that E_1 can never be larger than E_0 (since S_{xx} is a sum of squares and hence necessarily positive provided that the x_i are not all equal, while S_{xy}^2 is also a square and hence non-negative). Thus use of the explanatory variable X will *always* reduce the residual sum of squares for the data used to fit the model, which seems to say that we will always do better by including X in the prediction. However, this result is illusory, because we are usually interested in predicting *future* values which have not yet arisen, and which have not therefore been incorporated into the data from which the model has been fitted. By definition, the parameter estimates in any model are the *best* ones that can be obtained (in regression, they are the ones which *minimize* the sum of squared residuals as argued above). Thus they will be influenced by any chance fluctuations in the data, and applying the fitted model to future data will always give poorer predictions than when it is applied to the data on which it was fitted.

In particular, it can be demonstrated that the reduction from E_0 to E_1 still takes place even if the X values are completely randomly generated and have no relationship with the Y values. Consequently, we can only say that fitting the regression is beneficial if the reduction from E_0 to E_1 is *sufficiently large* for chance fluctuations to be overridden and for future predictions to be genuinely improved. Deciding whether this reduction is sufficiently large can be done through an *analysis of variance*. This is a formal method for partitioning the total variation in the sample of y_i values into the variation explained by the regression line and the residual variation about the fitted line. We have already seen above that

$$E_0 = \sum_{i=1}^{n} (y_i - \bar{y})^2 = S_{yy}$$

and

$$E_1 = \sum_{i=1}^{n} (y_i - \hat{\beta}_0 + \hat{\beta}_1[x_i - \bar{x}])^2 = S_{yy} - \frac{S_{xy}^2}{S_{xx}}.$$

Thus

$$E_0 - E_1 = \frac{S_{xy}^2}{S_{xx}}.$$

Now just below the definition of E_0 on page 58 we had

$$\hat{y}_i = \hat{\beta}_0 + \hat{\beta}_1(x_i - \bar{x}) = \bar{y} + \frac{S_{xy}}{S_{xx}}(x_i - \bar{x}),$$

so that on rearrangement of terms we have

$$\frac{S_{xy}}{S_{xx}}(x_i - \bar{x}) = \hat{\beta}_0 + \hat{\beta}_1(x_i - \bar{x}) - \bar{y} = \hat{y}_i - \bar{y}.$$

Thus, squaring and summing over i produces

$$\frac{S_{xy}^2}{S_{xx}^2}\sum_{i=1}^{n}(x_i - \bar{x})^2 = \sum_{i=1}^{n}(\hat{\beta}_0 + \hat{\beta}_1[x_i - \bar{x}] - \bar{y})^2 = \sum_{i=1}^{n}(\hat{y}_i - \bar{y})^2.$$

But $S_{xx} = \sum_{i=1}^{n}(x_i - \bar{x})^2$, so it follows that

$$E_0 - E_1 = \sum_{i=1}^{n}(\hat{\beta}_0 + \hat{\beta}_1[x_i - \bar{x}] - \bar{y})^2 = \sum_{i=1}^{n}(\hat{y}_i - \bar{y})^2.$$

This quantity represents the variation *accounted for* by the relationship between Y and X; it is termed the *regression sum of squares*. In similar vein E_1 represents the residual variation *about* the regression line so is termed the *residual sum of squares*, while E_0 is clearly the *total sum of squares* (about the sample mean). The fundamental analysis of variance identity thus states that the total sum of squares equals the regression sum of squares plus the residual sum of squares.

Associated with each sum of squares are degrees of freedom. The total sum of squares represents n deviations from the sample mean $\bar{y}$, which is an estimate of the population mean μ, so has $n - 1$ degrees of freedom (since the constraint $\sum(y_i - \bar{y}) = 0$ means that given any $n - 1$ values of Y along with the mean $\bar{y}$, the nth can be deduced). The residual sum of squares represents n deviations from the regression line which has two estimated parameters $\hat{\beta}_0$ and $\hat{\beta}_1$, so for similar reasons has $n - 2$ degrees of freedom. Thus one degree of freedom is left over for the regression sum of squares.

Dividing any sum of squares by its degrees of freedom yields the corresponding *mean square*. The residual mean square is the average variation around the regression line, so provides an estimate of σ^2, the variance of the departure terms ϵ_i in the model (3.1). This is *always* the case, even for the more complicated regression situations considered later. For this reason, the residual mean square is usually denoted by s^2. (However, the use of s^2 in this context should be carefully distinguished from its common usage as the sample variance $\frac{1}{n-1}\sum(y_i - \bar{y})^2$ of the y_i.)

Further theory establishes that *if there is no relationship between Y and X* then the regression sum of squares just represents random variation so the regression mean square is another, independent, estimate of σ^2, but if there *is* a genuine relationship between Y and X then the regression sum of squares contains this systematic element as well as random variation so will be much 'larger'. Thus if we calculate the ratio of the regression to residual mean squares, the value of this ratio will be close to 1 if there is no effective relationship between Y and X but it will be 'large' if there is an effective relationship.

All these features are concisely embodied in an *analysis of variance table* (generally abbreviated to the acronym ANOVA), which is set out as follows.

Source of Variation	Sum of Squares	Degrees of Freedom	Mean Square	F-Ratio
Regression	$SS_R = \frac{S_{xy}^2}{S_{xx}}$	1	$MS_R = SS_R$	$F = \frac{MS_R}{MS_E}$
Residual	$SS_E = S_{yy} - \frac{S_{xy}^2}{S_{xx}}$	$n - 2$	$MS_E = \frac{SS_E}{n-2}$	
Total	S_{yy}	$n - 1$		

The only remaining question is how to judge whether the ratio F is large enough to declare that there *is* an effective relationship between Y and X and that using X will be beneficial in predicting Y. The essential theory here rests on the fact that sums of squares of normal variables have chi-squared distributions, as do estimators of variance in general (Section 2.3). Thus the residual sum of squares, SS_E above, divided by σ^2, always has a chi-squared distribution on $n - 2$ degrees of freedom. Moreover, *if there is no true relationship between Y and X* then the regression sum of squares, SS_R above, also divided by σ^2, has an independent chi-squared distribution on 1 degree of freedom. Thus, in this case, the ratio F has an F-distribution on 1 and $n-2$ degrees of freedom (Section 2.3). However, if there *is* a true relationship between Y and X then the distribution of SS_R/σ^2 is a *non-central* chi-squared on 1 degree of freedom, which is much 'bigger' than an ordinary (central) chi-squared, so the ratio F will have much larger values than those from an $F_{1,n-2}$ distribution. We thus need to find the upper $100\alpha\%$ critical value $F_{1,n-2,\alpha}$ of the $F_{1,n-2}$ distribution from tables, and if our calculated ratio F exceeds $F_{1,n-2,\alpha}$ then we deem the regression to be 'significant' at the $100\alpha\%$ level. In this case, use of X *will* improve predictions of Y. However, if the calculated ratio F is less than $F_{1,n-2,\alpha}$ then the regression is 'not significant' at the chosen level and use of X *won't* improve predictions of Y.

A summary measure of the quality of the fit of the regression line to the data is given by the quantity R^2, defined to be the ratio of the regression sum of squares (i.e. S_{xy}^2/S_{xx}) to the total sum of squares (i.e. S_{yy}). This quantity, known as the *coefficient of determination* represents the proportion of the total sum of squares 'explained by the regression'. Thus the larger is R^2, the better is the fit of the straight line to the data. In fact, from the quantities given above, we see that

$$R^2 = S_{xy}^2/(S_{xx}S_{yy}),$$

which is just the square of the correlation between X and Y (Section 2.2). It can alternatively be established that R^2 is the square of the correlation between the y_i and their predicted values $\hat{y}_i$ from the model (and this latter fact remains true for the more complicated models considered later). Values of R^2 thus lie between 0 and 1 (although in computer software they are often expressed as percentages rather than proportions), and so a value close to 1 is required before the fit can be described as 'good'.

One cautionary note should be sounded here. It is important not to confuse the F-test in the analysis of variance with the measure of quality of regression given by R^2. The former simply indicates whether there is evidence of a linear relationship between Y and X, and *not* how good that relationship is (which is the role of R^2). It is entirely possible to have a very highly significant regression but with a relatively low value of R^2. Such an outcome would correspond to data for which the fitted regresion line was steep (i.e. slope considerably different from zero), so that changing X induced a large change in Y, but the points were widely scattered about the fitted line, so that there was still considerable error in predicting Y from this line.

Example 3.2. For the age and cholesterol level values in Table 3.1, we can add the summary value $S_{yy} = 13.920$ to those given in Example 3.1. We thus find

$$SS_R = S_{xy}^2/S_{xx} = 217.858^2/4139.833 = 11.465$$

and hence

$$SS_E = 13.920 - 11.465 = 2.455.$$

(Note that it is always easiest to obtain the residual sum of squares by subtraction in this way.) There are 24 individuals in the sample, so the numbers of degrees of freedom are 1 for regression and 22 for residual, whence $MS_E = SS_E/22 = 0.112$ and $F = 11.465/0.112 = 102.75$. The ANOVA table thus becomes:

Source	S.S.	d.f.	M.S.	F-Ratio
Regression	11.465	1	11.465	102.75
Residual	2.455	22	0.112	
Total	13.920	23		

The upper 0.1 percent point (i.e. $\alpha = 0.001$) of the F-distribution on 1 and 22 degrees of freedom is 14.0, and the calculated ratio 102.75 is hugely in excess of this. Consequently the regression is highly significant, and age should therefore substantially improve the prediction of cholesterol level. The quality of fit of regression is measured by $R^2 = 11.465/13.920 = 0.824$, indicating a high correlation between observed y_i and fitted $\hat{y}_i$.

3.2.4   Inferences about the regression parameters

Of frequent interest to the investigator is the making of statements about the parameters β_0, β_1 in the model (3.2), as these parameters summarize the assumed relationship between X and Y. The slope parameter β_1 is of particular interest, as this parameter quantifies the direct influence of X on Y. Now β_1 is a population parameter, while $\hat{\beta}_1$ is our best estimate of it in a particular situation. This estimate is of course just a sample statistic in the sense discussed in Section 2.4, so its value will vary when calculated for repeated samples of (x_i, y_i) pairs from the same population.

To determine the behaviour of $\hat{\beta}_1$ over such repeated samples we need its sampling distribution, and this can be obtained quite easily using theory in Chapter 2 allied to assumptions of the model. From this model, (3.2), we see that the y_i are independently distributed, the mean and variance of y_i being $\beta_0 + \beta_1(x_i - \bar{x})$ and σ^2 respectively. Moreover, for conducting inferences, it is essential to assume normality for the y_i. Now $\hat{\beta}_1$ can be written as $(\sum[x_i - \bar{x}]y_i)/(\sum[x_i - \bar{x}]^2)$ from direct solution of the normal equations above (3.3), and this is of the form $\sum l_i y_i$ where $l_i = [x_i - \bar{x}]/(\sum[x_i - \bar{x}]^2)$. Thus $\hat{\beta}_1$ is a linear combination of independent normal variables (the y_i), so by the results of Section 2.4 we can deduce that its sampling distribution is also normal, with mean $\sum l_i E(y_i)$ and variance $\sigma^2(\sum l_i^2)$.

Now $E(y_i) = \beta_0 + \beta_1(x_i - \bar{x})$, so

$$\sum l_i E(y_i) = \sum \left\{ \frac{[x_i - \bar{x}][\beta_0 + \beta_1(x_i - \bar{x})]}{\sum[x_i - \bar{x}]^2} \right\}$$

$$= \beta_0 \frac{\sum[x_i - \bar{x}]}{\sum[x_i - \bar{x}]^2} + \beta_1 \frac{\sum[x_i - \bar{x}]^2}{\sum[x_i - \bar{x}]^2}$$

$$= \beta_1,$$

the first term in the second line disappearing because the sum in the numerator is always zero, and the second fraction in this line cancelling to 1. Furthermore,

$$\sum l_i^2 = \sum \left\{ [x_i - \bar{x}]/\left(\sum[x_i - \bar{x}]^2\right) \right\}^2$$

$$= \left(\sum[x_i - \bar{x}]^2\right)/\left(\sum[x_i - \bar{x}]^2\right)^2$$

$$= 1/\left(\sum[x_i - \bar{x}]^2\right).$$

Thus the sampling distribution of $\hat{\beta}_1$ is normal, with mean β_1 (so that $\hat{\beta}_1$ is an *unbiased* estimator of β_1) and variance σ^2/S_{xx}. Remembering that σ^2 is estimated by the residual mean square in the analysis of variance table above, and denoting this estimate by s^2, it follows that the estimated standard error of $\hat{\beta}_1$ is $s/\sqrt{S_{xx}}$. (In practice, of course, σ^2 *always* has to be estimated, so henceforth we will assume the word 'estimated' when referring to a standard error and not mention it explicitly each time.)

Standard theory for confidence intervals (Section 2.5.2) thus shows that a $100(1 - \alpha)\%$ confidence interval for the 'true' value β_1 is given by

$$\hat{\beta}_1 \pm t_{n-2,\alpha/2} \frac{s}{\sqrt{S_{xx}}}, \tag{3.5}$$

where $t_{n-2,\alpha/2}$ is the upper $\alpha/2$ critical value of the t-distribution on $n - 2$ degrees of freedom (the number on which the estimate s^2 is based). Furthermore, to test the hypothesis that the slope of the regression line equals some specified value β_{10}, we calculate the test statistic

$$t = \frac{\hat{\beta}_1 - \beta_{10}}{s/\sqrt{S_{xx}}} \tag{3.6}$$

and refer the calculated value to the t-distribution on $n - 2$ degrees of freedom to determine its significance level.

There are two features of the above results that merit further mention. The first is that the standard error of $\hat{\beta}_1$ depends on S_{xx}: the greater the spread of the x values, the smaller is this standard error and hence the better defined is the line. The second feature is that we can use the above test for the hypothesis that there is no dependence of y on x, as this hypothesis is just the same as saying that the slope of the regression line is zero. Thus we can set $\beta_{10} = 0$ in the test statistic t of (3.6). However, this test is exactly equivalent to the ANOVA F-test previously discussed: the calculated value of the t statistic is equal to the square root of the previously obtained ratio of mean squares, and the square of the t-distribution on $n - 2$ degrees of freedom is identical to the F-distribution on 1 and $n - 2$ degrees of freedom.

Note finally that, since $\hat{\beta}_0 = \bar{y}$, inferences about the other unknown parameter β_0 are conducted by the usual methods based on $\bar{y}$. Standard results show that $\bar{y}$ has a normal distribution with mean β_0 and variance σ^2/n, which enables tests of hypothesis to be conducted and confidence intervals to be found. Moreover, using model (3.2) instead of (3.1) introduces the additional benefit that the parameters are *orthogonal* (see Section 3.3.3 below), so that $\hat{\beta}_0$ and $\hat{\beta}_1$ are *uncorrelated*. This means that inferences can be conducted on both parameters without fear of entanglement of effects.

Example 3.3. Continuing with the age and cholesterol data, we see from the residual mean square of the ANOVA in Example 3.2 that $s^2 = 0.112$ (so that $s = \sqrt{0.112} = 0.335$), and from Example 3.1 that $S_{xx} = 4139.833$. Also, $n = 24$ and the upper 2.5% critical value of the t-distribution on $24 - 2 = 22$ degrees of freedom is 2.075. Thus the standard errors of $\hat{\beta}_0$ and $\hat{\beta}_1$ are

$0.335/24 = 0.014$ and $0.335/\sqrt{4139.833} = 0.0052$ respectively. The 95% confidence interval for β_0 is thus

$$3.354 \pm 2.075 \times 0.014 = (3.325, 3.383)$$

while that for β_1 is

$$0.0526 \pm 2.075 \times 0.0052 = (0.042, 0.063).$$

To test the null hypothesis that $\beta_1 = 0$, we calculate the test statistic

$$t = 0.0526 \div 0.0052 = 10.13,$$

and refer this to the t-distribution on 22 degrees of freedom. The calculated value is considerably greater than the above 5% (two-tailed) critical value of 2.075, so we conclude that the result is highly significant and reject the null hypothesis in favour of the alternative that $\beta_1 \neq 0$. The same conclusion was reached in the ANOVA of Example 3.2; we see that the square of the above t statistic is $10.13^2 = 102.62$, which equals the ANOVA F-ratio of 100.75 to within rounding errors.

3.2.5 Prediction

We can distinguish two separate cases. First, we might wish to predict the *average* response of all individuals whose X values are x_0. Directly from (3.2), we see that this average is predicted as

$$\hat{\mu} = \hat{\beta}_0 + \hat{\beta}_1 (x_0 - \overline{x}).$$

However, this prediction is subject to sampling fluctuations in the estimates of the parameters, so it is only useful if accompanied by a standard error to quantify uncertainty. Since $\hat{\beta}_0$ and $\hat{\beta}_1$ are uncorrelated (see above), then the variance of the prediction is given by $\text{var}(\hat{\beta}_0) + (x_0 - \overline{x})^2 \, \text{var}(\hat{\beta}_1)$. Substitution of the variances from the preceding section into this expression, we find that

$$\sigma^2 \left(\frac{1}{n} + \frac{(x_0 - \overline{x})^2}{S_{xx}} \right)$$

is the prediction variance. Estimating σ^2 by s^2 as usual, and taking square roots, we thus see that the standard error of $\hat{\mu}$ is

$$S_M = s \sqrt{\left\{ \frac{1}{n} + \frac{(x_0 - \overline{x})^2}{S_{xx}} \right\}}.$$

The second case is when we wish to predict the response for a specified *individual* observation at $x = x_0$. The difference now is that we have to add

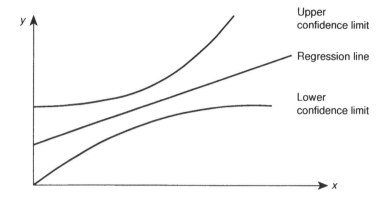

Fig. 3.3 *Illustration of confidence limits for prediction from regression line.*

the individual's contribution ϵ_i to the predicted mean response at $x = x_0$. Of course the best single prediction of ϵ_i is its mean (zero), so the point estimate is the same as above, i.e.

$$\hat{y}_i = \hat{\beta}_0 + \hat{\beta}_1(x_0 - \bar{x}).$$

However, the standard error will change, as we must now add the extra variation due to the individual value on top of that due to the prediction of the mean. The variance of ϵ_i is σ^2, and ϵ_i is independent of all previous observations. Hence the variance of the prediction is now $\sigma^2\left(1 + \frac{1}{n} + \frac{(x_0 - \bar{x})^2}{S_{xx}}\right)$ and the standard error is thus

$$S_I = s\sqrt{\left\{1 + \frac{1}{n} + \frac{(x_0 - \bar{x})^2}{S_{xx}}\right\}}.$$

The important point to note is that both standard errors above depend on the value x_0 of the explanatory variable at which prediction is required, and, more specifically, that the variances increase quadratically as x_0 moves away from the mean $\bar{x}$ of the explanatory variable values for the data on which the regression was fitted. Confidence in prediction is best at the centre $\bar{x}$ of the data set, but the confidence interval for either type of predicted value is curved and prediction becomes increasingly poorer as we move away from the centre. The situation is illustrated in Figure 3.3.

Thus prediction on the basis of a fitted regression relationship needs to be undertaken with care. Particularly dangerous is extrapolation beyond the range of the data used to fit the relationship. Not only is there no evidence that the fitted relationship remains valid beyond the range of the observed data but, even if it does remain valid, the resultant standard errors may be so large as to render the results practically useless. Care and thought in the use of regression cannot be overstressed!

Example 3.4. Suppose that the next patient to arrive is aged 40. This patient's predicted cholesterol level would be $\hat{y}_i = 1.281 + 0.0526 \times 40 = 3.385$, and the standard error of this prediction is

$$S_I = (0.335)\sqrt{\left\{1 + \frac{1}{24} + \frac{(40 - 39.417)^2}{4139.833}\right\}} = 0.349.$$

Thus, a 95% confidence interval for this prediction is $3.385 \pm 2.075 \times 0.349 = (2.661, 4.109)$; compare the width of this interval (1.45) with that (3.18) obtained at the start of Section 3.1 when age was *not* used in the prediction. Note also that prediction of *mean* cholesterol level for all 40-year olds is much more precise, with a standard error of

$$S_M = (0.335)\sqrt{\left\{\frac{1}{24} + \frac{(40 - 39.417)^2}{4139.833}\right\}} = 0.068.$$

The 95% confidence interval for this mean level is thus $(3.243, 3.527)$.

3.2.6 Checking model assumptions

All the foregoing theory concerning parameter estimation, confidence interval construction and hypothesis testing has required the model assumptions given in equation (3.1), and normality of the departures ϵ_i is also necessary for the inferential aspects. All these procedures will therefore be seriously compromised if the data do not accord with the assumptions of the model. Some assumptions, such as linearity of relationship between Y and X should be verified by plotting the data in a scattergram *before* any analysis is conducted, but other assumptions can only be checked when the analysis has been completed. In all cases, the aim is to avoid presenting results which appear (from computer output, say) excellent but which in fact are wrong in some systematic fashion.

A simple but very effective way of checking model assumptions is to inspect appropriate plots of the residuals $y_i - \hat{y}_i$, $i = 1, \ldots, n$. Since the simple regression model of this section is a special case of the multiple regression model to be considered in the next section, we defer discussion of underlying theory and methods for examination of residuals until we have considered the latter model (see Section 3.5). For the present we just note that if all model assumptions have been satisfied then the residuals should behave approximately like a random sample from a normal distribution centred at zero, while if some or all assumptions are violated then there will be some systematic warning pattern evident in an appropriate plot of the residuals. In particular, to check

(i) *linearity*, plot residuals (vertical axis) against x_i (horizontal axis): any non-linearity is shown up by a systematic non-linear pattern in the plot;

(ii) *normality*, arrange the residuals in increasing order from smallest to largest, obtain from normal tables the values z_i satisfying $\Phi(z_i) = (i - \frac{1}{2})/n$ for

$i = 1, \ldots, n$ (where $\Phi(\cdot)$ is the cumulative normal distribution function defined in Section 2.3.2), and plot the ranked residuals against the (ranked) z_i: departures from approximate linearity of the plot indicate departures from normality of data;

(iii) *homogeneity of variance*, plot residuals (vertical axis) against x_i, y_i or $\hat{y}_i$ (horizontal axis): equal scatter of points about the horizontal axis indicates homogeneity of variance, but unequal scatter implies that variance depends in some way on the variable plotted on the horizontal axis;

(iv) *independence*, plot residuals (vertical axis) against order in which the corresponding observations were made (horizontal axis): random scatter of points about horizontal axis indicates independence, but clusters of successive positive or negative residuals suggest the presence of *serial correlation* (i.e. observations taken successively are correlated).

The possible remedies for each type of violation will be discussed in Section 3.5, following the general discussion of tests of assumptions.

3.3 Multiple regression

3.3.1 Introduction and motivation

Consider the following example. Twenty-five observations were taken on two variables, the average atmospheric temperature in degrees Fahrenheit (X) and the number of pounds of steam (Y) used per month, at various intervals in a steam plant at a large industrial concern. The data are given in columns 2 and 3 of Table 3.2. A predictive equation was required for the dependent variable Y, so the linear regression $\hat{y}_i = 13.623 - 0.0798x_i$ was derived from the data. This gave rise to the following ANOVA table:

Source	S.S.	d.f.	M.S.	F-Ratio
Regression	45.5924	1	45.5924	57.54
Residual	18.2234	23	0.7923	
Total	63.8158	24		

A plot of the data and fitted line are shown in Figure 3.4, while the fitted values $\hat{y}_i$ and residuals $y_i - \hat{y}_i$ are given in columns 5 and 7 of Table 3.2.

The ANOVA F-ratio of 57.54 is highly significant when compared with critical values from the $F_{1,23}$ distribution, showing that the atmospheric temperature is a useful predictive variable for pounds of steam used in the plant. Figure 3.4 confirms this conclusion, as there is a very clear (negative) linear relationship between Y and X, so that knowledge of X should improve prediction of Y. However, Figure 3.4 also shows there to be quite wide scatter of points around the regression line, while from the ANOVA table we calculate $R^2 = 45.5924/63.8158 = 0.714$. There is still room for further improvement,

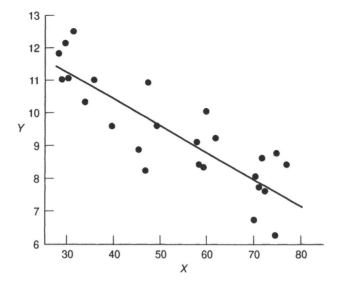

Fig. 3.4 *Scatter diagram and fitted regression line of Y on X for the data of Table 3.2.*

therefore, and this will come about if we can find an extra variable which has a relationship with the *residuals* from the regression of Y on X. Including this extra variable will then 'mop up' some of the residual variablity remaining after X has been used and the precision of prediction will thereby be increased.

Now it was observed that the number of days on which the plant was in operation varies from month to month, and the number of days on which the plant operates will clearly have some bearing on the number of pounds of steam used per month. Denoting the number of days in operation per month by Z, we might suppose that Z as well as X should be useful in prediction of Y. The values of Z corresponding to the previous observations made at the plant are given in column 4 of Table 3.2. Thus we ask whether these values are useful in predicting the *residuals* $y_i - \hat{y}_i$ from the previous regression.

However, since we have already used X in predicting Y, and since in general X may also have an influence on Z, we must first remove this influence before adding Z to the predictive equation. Thus, as a preliminary, we must regress Z on X and calculate the residuals $z_i - \hat{z}_i$. These residuals indicate what is left over after removing the effect of X, so they are the correct quantities for using in the prediction of $y_i - \hat{y}_i$. Regression of Z on X yields the relationship $\hat{z}_i = 22.169 - 0.037x_i$, and the fitted values $\hat{z}_i$ and residuals $z_i - \hat{z}_i$ from this regression are given in columns 6, 8 of Table 3.2.

Thus, finally, we ask whether the variation in the $y_i - \hat{y}_i$ is 'explainable' by variation in the $z_i - \hat{z}_i$, and to do this we regress the former residuals on

Table 3.2 *Observations (serial numbers i) on temperature (x_i), pounds of steam used per month (y_i) and number of operating days per month (z_i) in a steam plant, together with fitted values and residuals for the regressions of y_i on x_i and z_i on x_i (Source: Draper and Smith, 1981).*

i	x_i	y_i	z_i	$\hat{y}_i$	$\hat{z}_i$	$y_i - \hat{y}_i$	$z_i - \hat{z}_i$
1	35.3	10.98	20	10.81	20.87	0.17	−0.87
2	29.7	11.13	20	11.25	21.08	−0.12	−1.08
3	30.8	12.51	23	11.17	21.04	1.34	1.96
4	58.8	8.40	20	8.93	20.01	−0.53	−0.01
5	61.4	9.27	21	8.72	19.92	0.55	1.08
6	71.3	8.73	22	7.93	19.55	0.80	2.45
7	74.4	6.36	11	7.68	19.44	−1.32	−8.44
8	76.7	8.50	23	7.50	19.36	1.00	3.64
9	70.7	7.82	21	7.98	19.58	−0.16	1.42
10	57.5	9.14	20	9.03	20.06	0.11	−0.06
11	46.4	8.24	20	9.92	20.47	−1.68	−0.47
12	28.9	12.19	21	11.32	21.11	0.87	−0.11
13	28.1	11.88	21	11.38	21.14	0.50	−0.14
14	39.1	9.57	19	10.50	20.73	−0.93	−1.73
15	46.8	10.94	23	9.89	20.45	1.05	2.55
16	48.5	9.58	20	9.75	20.39	−0.17	−0.39
17	59.3	10.09	22	8.89	19.99	1.20	2.01
18	70.0	8.11	22	8.03	19.60	0.08	2.40
19	70.0	6.83	11	8.03	19.60	−1.20	−8.60
20	74.5	8.88	23	7.68	19.44	1.20	3.56
21	72.1	7.68	20	7.87	19.53	−0.19	0.47
22	58.1	8.47	21	8.98	20.04	−0.51	0.96
23	44.6	8.86	20	10.06	20.53	−1.20	−0.53
24	33.4	10.36	20	10.96	20.94	−0.60	−0.94
25	28.6	11.08	22	11.34	21.12	−0.26	0.88

the latter ones. When we carry out this regression we find the estimate of the intercept to be zero (as both sets of residuals have mean zero) and the estimate of the slope to be 0.203. Thus the predicted value of $y_i - \hat{y}_i$ is

$$\widehat{y_i - \hat{y}_i} = 0.203(z_i - \hat{z}_i), \tag{3.7}$$

while from the previous regressions we had $\hat{y}_i = 13.632 - 0.0798x_i$ and $\hat{z}_i = 22.169 - 0.037x_i$. Thus, substituting for $\hat{y}_i, \hat{z}_i$ in (3.7) and tidying up the resulting equation, we obtain $9.131 - 0.072x_i + 0.203z_i$ as the prediction of y_i when both X and Z are treated as explanatory variables. The correlation between the actual y_i and their predictions from this model is 0.921. The squared correlation is thus 0.85, which represents a substantial improvement over the R^2 of 0.714 obtained when X was the single predictor variable. Thus we can say that addition of Z has indeed been beneficial.

The process used above to incorporate this variable into the prediction has been rather tortuous (and has only been done in this way to illustrate the logical reasoning behind the modelling). In practice we would take a much more direct route, and this is now described.

3.3.2 The model

A *multiple regression* model is a model in which a *response variable* Y is linked to p (≥ 1) *explanatory (or regressor) variables* $X_1, X_2, \ldots, X_p$ and a random *departure* term. The model is *linear* if it is linear *in the parameters*, and the link between explanatory variables and departure term is most commonly an additive one (see Section 1.5).

Thus if $(y_i, x_{i1}, x_{i2}, \ldots, x_{ip})$ is the set of observations on the ith individual in a sample, then a multiple linear regression model is given by

$$y_i = \beta_0 + \beta_1 x_{i1} + \beta_2 x_{i2} + \cdots + \beta_p x_{ip} + \epsilon_i \qquad (i = 1, \ldots, n), \qquad (3.8)$$

where the departure terms ϵ_i are independent random variables each having mean zero and variance σ^2. As with simple regression, the explanatory variables are taken to be *fixed* and not random variables. By implication, the only random variable is the response variable Y, and the additional assumption of normality is necessary if any inferences are to be conducted on the β_i. The latter parameters are known as the *(partial) regression coefficients*.

It is convenient to define a dummy variable X_0 which takes the value 1 always (so that $x_{i0} = 1$ for all i), in which case the above model can be written

$$y_i = \beta_0 x_{i0} + \beta_1 x_{i1} + \cdots + \beta_p x_{ip} + \epsilon_i = \sum_{j=0}^{p} \beta_j x_{ij} + \epsilon_i \qquad (i = 1, \ldots, n). \ (3.9)$$

A more compact description, however, results from expressing the model in matrix and vector terminology. As such an expression leads to the best computational forms of all relevant formulae and quantities, this is the one we shall adopt and use in the following development. (It should be noted that it is of course entirely possible to avoid vectors and matrices in the derivation and statement of all required formulae, but such a scalar approach leads to very lengthy and somewhat indigestible formulae.)

To rewrite the model (3.9) in matrix form, we collect together all the observations on the response variable into the $(n \times 1)$ vector

$$\mathbf{y} = (y_1, y_2, \ldots, y_n)',$$

all the regression parameters into the $([p + 1] \times 1)$ vector

$$\boldsymbol{\beta} = (\beta_0, \beta_1, \ldots, \beta_p)',$$

and all the departure terms into the $(n \times 1)$ vector

$$\epsilon = (\epsilon_1, \epsilon_2, \ldots \epsilon_n)'.$$

where the prime denotes transpose. (It should be stressed that vectors here are always *column* vectors, so require the transpose symbol when written in rows as above.) Finally, we put all the observations on the explanatory variables in systematic fashion into an $(n \times [p+1])$ matrix X. The values for the ith individual, preceded by the value 1 (to match the constant β_0 of the model) constitute the ith row of this matrix. The matrix thus has the following form:

$$X = \begin{pmatrix} 1 & x_{11} & x_{12} & \cdots & x_{1p} \\ 1 & x_{21} & x_{22} & \cdots & x_{2p} \\ \vdots & \vdots & \vdots & & \vdots \\ 1 & x_{n1} & x_{n2} & \cdots & x_{np} \end{pmatrix}.$$

Writing the departure terms ϵ_i all together in a vector in this way produces a *random vector* ϵ which has a *multivariate* distribution, so we need a brief summary of the main features of multivariate distributions. Let $Z = (Z_1, Z_2, \ldots, Z_n)'$ be a random vector that has a multivariate distribution. The *mean vector* μ of this distribution is just the n-vector whose elements are the means of the individual Z_i, so that we can write $E(Z) = \mu$. The *variance–covariance* (or, more simply, the *covariance*) matrix Σ of this distribution is the $(n \times n)$ matrix which has the variances of the Z_i down its principal (i.e. left-to-right) diagonal and the covariances between pairs Z_i, Z_j $(i \neq j)$ in its off-diagonal (i, j) positions. Writing the covariance matrix as $\mathrm{cov}(Z)$, elementary matrix operations easily establish the results:

(i) $\mathrm{cov}(Z) = E(ZZ') - E(Z)E(Z')$;
(ii) $E(AZ + b) = A\mu + b$; and
(iii) $\mathrm{cov}(AZ + b) = A\Sigma A'$

for any constant matrix A and constant vector b.

Returning to (3.9), and putting all the above constituents together, we arrive at the following matrix form of the multiple regression model:

$$y = X\beta + \epsilon \tag{3.10}$$

where ϵ has mean vector $\mathbf{0}$ and covariance matrix $\sigma^2 I$.

Using the above properties of multivariate distributions, it immediately follows that $E(Y) = X\beta$ and $\mathrm{cov}(Y) = \sigma^2 I$. The form (3.10) is known as the *general linear model* and finds extensive use in many different contexts. For the present, however, we concentrate on its use in multiple regression.

Example 3.5. Expressing the example of Section 3.3.1 in the above form, we would denote the average atmospheric temperature by X_1, the number of operating days per month by X_2, and leave the pounds of steam per month as Y.

Then we have $n = 25$, $p = 2$,

$$y = (10.98, 11.13, 12.51, \ldots, 11.08)',$$

$$X = \begin{pmatrix} 1 & 35.3 & 20 \\ 1 & 29.7 & 20 \\ 1 & 30.8 & 23 \\ \vdots & \vdots & \vdots \\ 1 & 28.6 & 22 \end{pmatrix}$$

$$\beta = (\beta_0, \beta_1, \beta_2)',$$

and

$$\epsilon = (\epsilon_1, \epsilon_2, \ldots, \epsilon_{25})'.$$

3.3.3 Fitting the model

As in the case of simple linear regression, we estimate the parameters β_j ($j = 0, 1, \ldots, p$) by the method of least squares. Thus our aim is to find the vector β that minimizes

$$\begin{aligned} S(\beta) &= \sum_{i=1}^{n} \epsilon_i^2 \\ &= \epsilon'\epsilon \\ &= (y - X\beta)'(y - X\beta) \\ &= y'y - 2\beta'X'y + \beta'X'X\beta. \end{aligned}$$

The usual way of proceeding is by differentiating S with respect to each of the β_j, setting the resulting expressions to zero and solving the ensuing system of simultaneous equations. An alternative route not involving calculus is to note that if β_0 is any value of β satisfying

$$X'X\beta_0 = X'y, \tag{3.11}$$

then $S(\beta_0) \leq S(\beta)$ for all β. This result can be established in the following steps. First, from the definition of S above we have

$$S(\beta) - S(\beta_0) = \beta'X'X\beta - 2\beta'X'y + 2\beta_0'X'y - \beta_0'X'X\beta_0.$$

Next, substituting for $X'y$ from (3.11), we obtain

$$S(\beta) - S(\beta_0) = \beta'X'X\beta - 2\beta'X'X\beta_0 + \beta_0'X'X\beta_0.$$

Finally, collecting terms together and tidying up we see that

$$S(\beta) - S(\beta_0) = [X(\beta - \beta_0)]'[X(\beta - \beta_0)],$$

which is a sum of squares so must be greater than or equal to zero. This establishes the result.

If two distinct vectors $\beta_0^{(1)}$ and $\beta_0^{(2)}$ satisfy (3.11), then by subtraction

$$X'X(\beta_0^{(1)} - \beta_0^{(2)}) = \mathbf{0}$$

and elementary linear algebra tells us that this can only be the case if the determinant of $X'X$ is zero. Thus if $\det(X'X) \neq 0$, then $X'X$ is non-singular and

$$\hat{\beta} = (X'X)^{-1}X'y \tag{3.12}$$

is the (unique) least squares estimator of β.

We can immediately note that this estimator is of the form Ay (where $A = (X'X)^{-1}X'$), so that from properties (ii) and (iii) in Section 3.3.2 we find after simplification that

(a) $E(\hat{\beta}) = \beta$, and
(b) $\text{cov}(\hat{\beta}) = \sigma^2(X'X)^{-1}$.

The diagonal elements of $\text{cov}(\hat{\beta})$ give the variances of the $\hat{\beta}_j$, while the off-diagonal elements give covariances between pairs $\hat{\beta}_j, \hat{\beta}_k$. The square roots of the diagonal elements are thus the *standard errors* of the $\hat{\beta}_j$.

In general, the inverse of a matrix such as $X'X$ is a complicated function of all its elements. However, if the off-diagonal elements of $X'X$ are all zero then so are all the off-diagonal elements of the inverse, while the diagonal elements of the inverse are then just the inverses of the original diagonal elements. In this case the $\hat{\beta}_j$ are mutually uncorrelated, and we say that the explanatory (i.e. regressor) variables are *orthogonal*. Additional benefits arise in subsequent analysis when the variables are orthogonal, as we show later.

Example 3.6. Return to the simple linear regression model of (3.2). This, of course, is a special case of the general linear model (3.10) in which $p = 1$,

$$y = (y_1, y_2, \ldots, y_n)',$$

$$X = \begin{pmatrix} 1 & x_1 - \overline{x} \\ 1 & x_2 - \overline{x} \\ 1 & x_3 - \overline{x} \\ \vdots & \vdots \\ 1 & x_n - \overline{x} \end{pmatrix},$$

$$\beta = (\beta_0, \beta_1)',$$

and

$$\epsilon = (\epsilon_1, \epsilon_2, \ldots, \epsilon_n)'.$$

Straightforward matrix transposition and multiplication thus yields

$$X'X = \begin{pmatrix} n & 0 \\ 0 & \sum_j (x_j - \bar{x})^2 \end{pmatrix} = \begin{pmatrix} n & 0 \\ 0 & S_{xx} \end{pmatrix},$$

from which we immediately obtain

$$(X'X)^{-1} = \begin{pmatrix} 1/n & 0 \\ 0 & 1/S_{xx} \end{pmatrix}.$$

Hence

$$\text{cov}(\hat{\beta}) = \sigma^2 (X'X)^{-1} = \begin{pmatrix} \sigma^2/n & 0 \\ 0 & \sigma^2/S_{xx} \end{pmatrix}.$$

Thus we see that $\hat{\beta}_0$ and $\hat{\beta}_1$ are uncorrelated, and the regressors (in this case, trivially, X_0 and X_1) are orthogonal, as was asserted in Section 3.2.4. Moreover, the variances of $\hat{\beta}_0$ and $\hat{\beta}_1$ from the diagonal elements of $\sigma^2 (X'X)^{-1}$ agree with the expressions given in Section 3.2.4. Finally,

$$X'y = \begin{pmatrix} \sum y_j \\ \sum (x_j - \bar{x}) y_j \end{pmatrix} = \begin{pmatrix} \sum y_j \\ S_{xy} \end{pmatrix},$$

so that

$$\hat{\beta} = (X'X)^{-1} X'y = \begin{pmatrix} \frac{1}{n} \sum y_j \\ S_{xy}/S_{xx} \end{pmatrix}.$$

Again, these expressions agree with those previously derived in (3.3).

Example 3.7. For a numerical example, return to the 'steam plant' data in general linear model form as outlined in Example 3.5. Straightforward matrix operations yield

$$X'X = \begin{pmatrix} 25.0 & 1315.0 & 506.0 \\ 1315.0 & 76323.4 & 26353.3 \\ 506.0 & 26353.3 & 10460.0 \end{pmatrix},$$

and

$$X'y = \begin{pmatrix} 235.6 \\ 11820.7 \\ 4831.6 \end{pmatrix}.$$

Inverting the former matrix, we have

$$(X'X)^{-1} = \begin{pmatrix} 2.77875 & -0.01124 & -0.10610 \\ -0.01124 & 0.00015 & 0.00018 \\ -0.10610 & 0.00018 & 0.00479 \end{pmatrix}$$

from which, finally, we obtain

$$\hat{\beta} = \begin{pmatrix} 2.77875 & -0.01124 & -0.10610 \\ -0.01124 & 0.00015 & 0.00018 \\ -0.10610 & 0.00018 & 0.00479 \end{pmatrix} \begin{pmatrix} 235.6 \\ 11820.7 \\ 4831.6 \end{pmatrix} = \begin{pmatrix} 9.1302 \\ -0.0724 \\ 0.2027 \end{pmatrix}.$$

This set of estimates agrees excellently with the ones previously derived in Section 3.3.1.

3.3.4 Assessing the regression

Having obtained the least squares estimator $\hat{\beta}$ as in (3.12), it follows from the model (3.10) that the vector of *fitted values* $\hat{y} = (\hat{y}_1, \ldots, \hat{y}_n)$ corresponding to the observed values $y' = (y_1, \ldots, y_n)$ is

$$\hat{y} = X\hat{\beta}$$
$$= X(X'X)^{-1}X'y,$$

on substituting for $\hat{\beta}$ from (3.12). Thus we can write

$$\hat{y} = Hy, \tag{3.13}$$

where $H = X(X'X)^{-1}X'$. The matrix H is known as the *hat* matrix (because it 'puts a hat on y').

The vector of *residuals* $e' = (e_1, \ldots, e_n)$ is thus given by

$$e = y - \hat{y}$$
$$= y - X\hat{\beta}$$
$$= y - Hy$$
$$= (I - H)y. \tag{3.14}$$

The total (corrected) sum of squares of the response variable is again $S_{yy} = \sum(y_i - \bar{y})^2$, which in vector terminology is $y'y - n\bar{y}^2$. Adding and subtracting the quantity $\hat{\beta}'X'y$, we thus see that

$$S_{yy} = (y'y - \hat{\beta}'X'y) + (\hat{\beta}'X'y - n\bar{y}^2).$$

Now from the definition of $\hat{y}$ above, we see that

$$\hat{y}'\hat{y} = \hat{\beta}X'X\hat{\beta},$$

so substituting for the second $\hat{\beta}$ from (3.12) we have

$$\hat{y}'\hat{y} = \hat{\beta}X'y.$$

Thus the two terms in the first pair of parentheses on the right-hand side of the equation for S_{yy} above constitute the residual sum of squares,

$$SS_E = \sum(y_i - \hat{y}_i)^2,$$

of the individual y_i about the fitted regression. By analogy with the simple regression case of Section 3.2.3, therefore, the two terms in the second pair of parentheses form the regression sum of squares,

$$SS_R = \hat{\beta}'X'y - n\bar{y}^2,$$

and the total sum of squares S_{yy} breaks down into the same fundamental analysis of variance identity as before.

However, since there are now p regressor variables in the model, the number of degrees of freedom for regression is also p. The number of degrees of freedom for S_{yy} is of course unchanged at $n - 1$, so the number of degrees of freedom remaining for residual is $n - p - 1$. Also, it can be shown that the expected value (with respect to repeated sampling) of SS_E is $(n - p - 1)\sigma^2$, so an unbiased estimator of σ^2 is given by $s^2 = SS_E/(n - p - 1)$ (i.e. the residual mean square). We can again summarize all these features in an ANOVA table:

Source of Variation	Sum of Squares	Degrees of Freedom	Mean Square
Regression	$SS_R = \hat{\beta}' X' y - n\bar{y}^2$	p	$MS_R = SS_R/p$
Residual	$SS_E = y'y - \hat{\beta}' X' y$	$n - p - 1$	$MS_E = SS_E/(n - p - 1)$
Total	S_{yy}	$n - 1$	

The question that now arises is whether there is any benefit to be gained from relating Y to the Xs or not. Formally, this question can be answered by testing the null hypothesis

$$H_0 \quad : \quad \beta_1 = \beta_2 = \cdots = \beta_p = 0$$

(i.e. none of the Xs has any relationship with Y), against the alternative

$$H_a \quad : \quad \text{at least one } \beta_j \neq 0.$$

To derive a test, it can be shown that SS_E/σ^2 always has a chi-squared distribution on $n - p - 1$ degrees of freedom, and that *if H_0 is true* then SS_R/σ^2 has an independent chi-squared distribution with p degrees of freedom. Thus, *if H_0 is true* then $\frac{SS_R/p}{SS_E/(n-p-1)} = \frac{MS_R}{MS_E}$ has an F-distribution on p and $n - p - 1$ degrees of freedom. On the other hand, if H_0 is not true then this ratio has a non-central F-distribution on the same numbers of degrees of freedom, and values from this distribution are much bigger. To conduct the test, therefore, we calculate the ratio of mean squares (as we did in simple regression) and compare the result with critical values from the $F_{p,n-p-1}$ distribution. If the calculated value exceeds the upper $100\alpha\%$ critical value, then we say that the test is significant at the $100\alpha\%$ level and the Xs *do* have an influence on Y. If, on the other hand, the calculated value is less than the critical value then the test is not significant at the given significance level, in which case there is no evidence that any of the Xs are useful in predicting Y.

A summary measure of the quality of the model for prediction of Y is again given by the ratio of the regression sum of squares to the total sum of squares, i.e. $R^2 = SS_R/S_{yy}$, and it can be shown that this quantity equals the square of the correlation between the observed y_i and the fitted $\hat{y}_i$.

Example 3.8. Return again to the 'steam plant' data. From the ANOVA table in Section 3.3.1 we see that $S_{yy} = 63.82$, while from the raw data of Table 3.2 we can calculate $\bar{y} = 9.4236$. Thus $n\bar{y}^2 = 25 \times 9.4236^2 = 2220.11$.

Next, from the calculations in Example 3.7 we have

$$\hat{\beta}' X' y = 9.1302 \times 235.6 - 0.0724 \times 11820.7 + 0.2027 \times 4831.6 = 2274.62,$$

so that

$$SS_R = 2274.62 - 2220.11 = 54.51$$

and

$$SS_E = 63.82 - 54.51 = 9.31.$$

We can thus draw up the ANOVA table:

Source	S.S.	d.f.	M.S.	F-Ratio
Regression	54.51	2	27.255	64.433
Residual	9.31	22	0.423	
Total	63.82	24		

The upper 5 percent point (i.e. $\alpha = 0.05$) of the F-distribution on 2 and 22 degrees of freedom is 3.44, while the upper 1 percent point is 5.72. The calculated ratio 64.433 is hugely in excess of both critical values, so the regression is highly significant and we conclude that at least one of the two X variables (temperature and number of operating days per month) is a useful explanatory variable for pounds of steam used per month in the plant. Moreover, $R^2 = 54.51 \div 63.82 = 0.854$, so the quality of the fitted model is good. Our estimate of σ^2 is now $s^2 = 0.423$, which may be contrasted with the value 0.792 when only temperature was included in the regression relationship (see the ANOVA table in Section 3.3.1).

3.3.5 Inferences about individual regression parameters

We saw in Section 3.3.3 that $E(\hat{\beta}) = \beta$ and $\text{cov}(\hat{\beta}) = \sigma^2(X'X)^{-1}$. It thus follows that

$$E(\hat{\beta}_j) = \beta_j \quad \text{and} \quad \text{var}(\hat{\beta}_j) = \sigma^2 c_{jj},$$

where c_{jj} is the element in the $(j+1, j+1)$th position of $(X'X)^{-1}$ for $j = 0, 1, \ldots, p$.

If we now assume normality of the departures ϵ_i in model (3.10), then it follows that the y_i are normally distributed and, since the least squares estimator (3.12) has linear combinations of the y_i as its elements, the $\hat{\beta}_j$ are also normally distributed (Section 2.4). In fact, for $j = 0, 1, \ldots, p$,

$$\hat{\beta}_j \sim N(\beta_j, \sigma^2 c_{jj}), \tag{3.15}$$

so that

$$\frac{\hat{\beta}_j - \beta_j}{\sqrt{\sigma^2 c_{jj}}} \sim N(0, 1), \tag{3.16}$$

and

$$\frac{\hat{\beta}_j - \beta_j}{\sqrt{s^2 c_{jj}}} \sim t_{n-p-1} \tag{3.17}$$

where $\sqrt{s^2 c_{jj}} = s\sqrt{c_{jj}}$ is the *standard error* of $\hat{\beta}_j$.

It is tempting to use the distribution in (3.17) to test whether or not X_j is 'important' in explaining the variation in Y. Thus to test $H_0 : \beta_j = 0$ (i.e. X_j is not important) against $H_a : \beta_j \neq 0$ (i.e. X_j is important) we might use the test statistic

$$T_j = \frac{\hat{\beta}_j}{\sqrt{s^2 c_{jj}}},$$

and reject H_0 in favour of H_a at the $100\alpha\%$ level if the computed value of T_j lies outside the interval $(-t_{\alpha/2,n-p-1}, t_{\alpha/2,n-p-1})$. It is perfectly permissible to conduct such a test, but caution should be exercised in the interpretation of the results as this is a test of the significance of X_j *in the presence of all the other variables*. Thus if $\hat{\beta}_j$ turns out to be not significantly different from zero it means that X_j does not contribute significantly to variation in Y *after the contributions of all the other explanatory variables have been taken into account*. It is thus entirely possible that X_j may become 'significant' if some of the other Xs are dropped from the model. This fact has considerable importance when selecting the 'best' set of Xs to include in the model, as we shall see in Section 3.4.

Similarly, a $100(1 - \alpha)\%$ confidence interval for β_j is given by

$$\hat{\beta}_j - t_{\alpha/2,n-p-1}s\sqrt{c_{ii}} \leq \beta_j \leq \hat{\beta}_j + t_{\alpha/2,n-p-1}s\sqrt{c_{ii}},$$

but the same cautionary note as above applies to the interpretation.

Example 3.9. For the 'steam plant' data, we had $s^2 = 0.423$ above, while from Example 3.7 we see that $c_{11} = 2.77875$. It thus follows that the standard error of $\hat{\beta}_0$ is $\sqrt{0.423 \times 2.77875} = 1.084$. In similar fashion we find the standard errors of $\hat{\beta}_1$ and $\hat{\beta}_2$ to be 0.008 and 0.046 respectively. Thus the calculated values of the test statistic given above are $T_0 = 8.42$, $T_1 = -9.05$ and $T_2 = 4.41$. These values all lie considerably outside the interval $(-2.82, 2.82)$, where 2.82 is the upper 0.5 percent critical value of the t_{22} distribution, so that all coefficients are significantly different from zero at the 1% level. This means that each term is important *when all the others are present in the model*.

3.3.6 Prediction

Consider predicting a future value of Y assuming that the corresponding regressor variables $X_1, X_2, \ldots, X_p$ are known to have the values $x_{a1}, x_{a2}, \ldots, x_{ap}$ respectively. As with the simple regression situation of Section 3.2.5, we can distinguish two cases: predicting the mean response $\hat{\mu}$ of all individuals with these specified values of the regressor variables, and predicting the response $\hat{y}$ of a single individual with these values of the regressor variables. As in the previous situation, the point estimates will clearly coincide for the two cases but the standard errors will differ.

First let us write $x'_a = (1, x_{a1}, x_{a2}, \ldots, x_{ap})$. Then it is evident that the two point estimates are

$$\hat{y} = \hat{\mu} = x'_a \hat{\beta}. \tag{3.18}$$

As regards their standard errors, in the same way as before it can be noted that the variability in $\hat{\mu}$ arises purely from sampling fluctuations in $\hat{\beta}$, while the variability in $\hat{y}$ has in addition the extra variability σ^2 caused by the individual's variance about the mean response.

Thus, from the basic result (iii) in Section 3.3.2 concerning covariance matrices of linear transformations, we have

$$\operatorname{var}(\hat{\mu}) = x'_a [\operatorname{cov}(\hat{\beta})] x_a = \sigma^2 x'_a (X'X)^{-1} x_a$$

and hence

$$\operatorname{var}(\hat{y}) = \sigma^2 \{1 + x'_a (X'X)^{-1} x_a\}.$$

Estimating σ^2 by the residual mean square s^2 and taking square roots we thus obtain

$$S_M = s \sqrt{x'_a (X'X)^{-1} x_a}$$

for the standard error of $\hat{\mu}$, and

$$S_I = s \sqrt{1 + x'_a (X'X)^{-1} x_a}$$

for the standard error of $\hat{y}$. Confidence intervals can then be constructed in the usual way (assuming normality of responses and remembering that the estimate s^2 is based on $n - p - 1$ degrees of freedom).

Example 3.10. Continuing with the 'steam plant' example, suppose that in the next month we aim to maintain a temperature of 50 degrees and operate the plant on 20 days. What is a 95% confidence interval for the number of pounds of steam that will be used in the plant during this month?

Here $x'_a = (1, 50, 20)$, so using the vectors and matrices given in Example 3.7, we have

$$\hat{y} = 1 \times 9.1302 - 50 \times 0.0724 + 20 \times 0.2027 = 9.5642$$

pounds of steam as the point estimate prediction, and

$$x'_a(X'X)^{-1} = (0.09475, -0.00014, -0.00130)$$

so that

$$x'_a(X'X)^{-1}x_a = 0.06175.$$

Thus, since $s^2 = 0.423$ from Example 3.8, then

$$S_I = \sqrt{0.423 \times 1.06175} = 0.670.$$

We have 22 degrees of freedom for the estimation of s^2, and the appropriate critical value from the t_{22} distribution for 95% confidence intervals is 2.07. The required interval is thus given by

$$(9.5642 - 2.07 \times 0.670, 9.5642 + 2.07 \times 0.670), \qquad \text{i.e. } (8.177, 10.951).$$

3.3.7 Some possible problems

It is evident from all the foregoing that the matrix $(X'X)^{-1}$ plays a very important role in multiple regression, as it is involved in not only the calculation of the regression coefficients but also in the estimation of their standard errors and the standard errors of any fitted or predicted values. All of these calculations will therefore break down if $X'X$ is singular, as the inverse does not exist in this case. There is a very simple pragmatic way round such a problem, however, as singularity of $X'X$ is generally caused by some linear dependence existing among the regressor variables. The problem can therefore be overcome by leaving out a sufficient number of these variables for the linear dependence to be removed, and it is generally understood that such action will be taken as necessary so that non-singularity of $X'X$ can be assumed.

Severe problems can nevertheless be encountered if $X'X$ *can* be inverted but is *nearly* singular, which will happen if there is an *approximate* linear relationship among some or all of the regressor variables. In this case we say that there is *multicollinearity* among the regressor variables. The main problems caused by multicollinearity are that

- some or all of the regression coefficients will have large standard errors (so are unreliable as estimators of the model parameters),
- there is instability in the fitted models (meaning that a small perturbation to an observation, or deletion of an observation from the data set will produce a very different fitted model),
- difficulties arise in variable selection (see Section 3.4 below).

It is thus important to recognize multicollinearity if it is present, so that remedial action can be taken.

Fortunately it is easy to obtain useful diagnostics for the condition. Since we are looking for approximate linear dependencies among the regressor variables, and since a 'well fitting' regression model is evidence of a close linear relationship (i.e. dependency) among a set of variables, then all we need to do is to regress each X_j on all of the other Xs. If there *is* an approximate dependency among the regressor variables then at least one of these regressions will fit well, i.e. have a large R^2 (close to 1). The X variable that is used as the response variable in such a regression can thus be very well predicted by the other X variables, so may be treated as redundant and removed from the model. This will generally stabilize the model-fitting process.

One further connection removes even the need to conduct a set of regressions in this way to detect multicollinearity. Let us write R_j^2 for the coefficient of determination when X_j is regressed on all the other Xs. Then it can be shown that the $(j + 1, j + 1)$th element $c_{j+1,j+1}$ of $(X'X)^{-1}$ is equal to $(1 - R_j^2)^{-1}$. Thus as R_j^2 approaches 1, $c_{j+1,j+1}$ becomes very large and provides a warning that multicollinearity is present through the near dependence of X_j on the other Xs. The diagonal elements $c_{j+1,j+1}$ are known as the *variance inflation factors* for the regression, and a useful rule of thumb is to suspect the presence of multicollinearity when any of these factors becomes larger than about 10 (i.e. when $R_j^2 > 0.9$ for some j).

A related problem occurs if we have a very disparate collection of regressor variables such that the units of measurement of different Xs vary considerably. Then the elements of $X'X$ will vary considerably in magnitude, and we may have very poor accuracy when computing $(X'X)^{-1}$. In this case, $X'X$ is said to be *ill-conditioned*. The way round this problem is to *scale* all the variables first, by setting

$$y_i' = \frac{y_i - \bar{y}}{\sqrt{S_{yy}}} \quad \text{and} \quad x_{ij}' = \frac{x_{ij} - \bar{x}_j}{\sqrt{S_{jj}}}$$

where $S_{yy} = \sum_i (y_i - \bar{y})^2$ and $S_{jj} = \sum_i (x_{ij} - \bar{x}_j)^2$, for $i = 1, \ldots, n$ and $j = 1, \ldots, p$, and then to fit the model

$$y_i' = \theta_1 x_{i1}' + \theta_2 x_{i2}' + \cdots + \theta_p x_{ip}' + \epsilon_i.$$

Scaling the variables in this way equalizes all the magnitudes of elements of the cross-product matrix, and maximizes the computational accuracy when inverting this matrix. The regression coefficients $\hat{\theta}_j$ computed from scaled regressor variables are often called (somewhat confusingly) the *beta coefficients*, and the original regression coefficients $\hat{\beta}_j$ can be recovered from the relationships

$$\hat{\beta}_j = \hat{\theta}_j \sqrt{\frac{S_{yy}}{S_{jj}}} \quad (j = 1, \ldots, p)$$

and

$$\hat{\beta}_0 = \bar{y} - \sum_j \hat{\beta}_j \bar{x}_j.$$

3.3.8 Some special cases

Model (3.9) is a general formulation in terms of a set of p regressor variables, and admits a number of special cases in which some or all of these regressor variables have a particular form. We give two of the most common such cases.

Polynomial models

Recollect from Section 1.5 that when we say a model is linear, we mean that it is linear *in the parameters* $\beta_0, \beta_1, \ldots, \beta_p$. Thus, providing this condition is satisfied, we can fit models that include powers or products of any of the regressors simply by redefining each power or product as a new regressor variable and using the standard theory developed above. For example,

(a) a second-degree polynomial model in one variable X, viz.

$$y_i = \beta_0 + \beta_1 x_i + \beta_2 x_i^2 + \epsilon_i,$$

can be re-expressed in the form

$$y_i = \beta_0 x_{i0} + \beta_1 x_{i1} + \beta_2 x_{i2} + \epsilon_i$$

by setting $x_{i0} = 1$, $x_{i1} = x_i$ and $x_{i2} = x_i^2$ for all $i = 1, \ldots, n$;

(b) a second-order response surface model in two variables X_1 and X_2, viz.

$$y_i = \beta_0 + \beta_1 x_{i1} + \beta_2 x_{i2} + \beta_3 x_{i1}^2 + \beta_4 x_{i2}^2 + \beta_5 x_{i1} x_{i2} + \epsilon_i,$$

can be re-expressed in the form

$$y_i = \beta_0 x_{i0} + \beta_1 x_{i1} + \beta_2 x_{i2} + \beta_3 x_{i3} + \beta_4 x_{i4} + \beta_5 x_{i5} + \epsilon_i$$

by setting $x_{i0} = 1$, $x_{i3} = x_{i1}^2$, $x_{i4} = x_{i2}^2$ and $x_{i5} = x_{i1} x_{i2}$ for all $i = 1, \ldots, n$;

and so on.

Although we should be able to fit such models by standard theory, note that high correlations will inevitably exist between powers such as X, X^2, X^3 etc. Also, powering rapidly increases the magnitudes of values. Hence as the order of model increases so will the danger of $X'X$ becoming ill-conditioned, and we are increasingly likely to encounter multicollinearity problems. Mean-centering is a useful device in such circumstances, i.e. use $z_i = (x_i - \bar{x})$ instead of x_i, fit a polynomial model in terms of z_i, and deduce the parameters for the polynomial in x_i from this fitted model. For example, if we wish to fit

$$y_i = \beta_0 + \beta_1 x_i + \beta_2 x_i^2 + \epsilon_i,$$

we fit instead

$$y_i = \gamma_0 + \gamma_1 z_i + \gamma_2 z_i^2 + \epsilon_i$$

where $z_i = x_i - \bar{x}$. The latter model is just

$$y_i = \gamma_0 + \gamma_1(x_i - \bar{x}) + \gamma_2(x_i - \bar{x})^2 + \epsilon_i,$$

so on expanding the squared term and collecting terms together we see that

$$\beta_0 = \gamma_0 - \gamma_1\bar{x} + \gamma_2\bar{x}^2,$$
$$\beta_1 = \gamma_1 - 2\gamma_2\bar{x},$$

and $\beta_2 = \gamma_2$. Thus estimates $\hat{\beta}_i$ are readily deduced from the $\hat{\gamma}_i$, and the latter are obtained with greater numerical stability because many of the correlations between regressors are greatly reduced by the mean-centering.

Example 3.11. A study was undertaken to determine the most economical date of planting of cotton in uplands regions of Bangladesh. Cotton yield is known to be directly related to the time of planting, and previous work suggested that the optimum time of planting was September. However, since heavy rains occasionally interfere with planting at this time, the study was designed to investigate the relationship of yield and time of planting over a three-month period commencing on 1 September. The data below give the day of planting after 1 September (X), and the yield of the cotton in tens of kilogrammes per hectare (Y).

x_i	1	16	31	46	61	76	91	106	121	136
y_i	17.39	17.74	16.02	14.34	13.38	9.78	7.38	6.09	4.26	3.92

A plot of yield against day shows a progressive decrease, but with a suggested 'point of inflection' around the 6–8 weeks mark. The decrease is thus (possibly) non-linear with time, and we might therefore try fitting a cubic model of the form

$$y_i = \beta_0 + \beta_1 x_i + \beta_2 x_i^2 + \beta_3 x_i^3 + \epsilon_i$$

for $i = 1, \ldots, 10$. Undue problems of multicollinearity were not encountered, so this model was fitted directly in raw form giving the following parameter estimates, standard errors and t-ratios:

Parameter	Estimate	St. Error	t-ratio
β_0	17.2579	0.5966	28.93
β_1	0.0670	0.0398	1.68
β_2	−0.0032	0.00069	−4.62
β_3	1.46×10^{-5}	3.3×10^{-6}	4.39

All parameter estimates except the linear coefficient are significant at the 1% level. In particular, the cubic coefficient is significant in the presence of the others, so we are justified in fitting the cubic model. (We examine this aspect in a little more detail later – see Example 3.13). The fact that the linear term is not significant in the presence of the others is of little consequence, as including a cubic term logically implies that we should include all the lower-order terms also. The coefficient of determination of the fitted cubic is $R^2 = 0.991$, indicating an excellent fit which gives very good prediction of yield over the range of experimentation. We might thus conclude that the latest viable time of planting should be at the point of inflection of the fitted cubic, which is easily shown to be at 68 days.

Models that incorporate factors

Regressor variables are usually quantitative (as has been the case in all examples so far), but sometimes it is necessary to incorporate *qualitative* variables or *factors* in the model. For example, we may have information on such variables as gender (with categories male/female), or type of soil (with categories sandy/stony/peaty), or season (with categories spring/summer/autumn/winter). Such information is easily incorporated in the model with the help of *indicator* or *dummy* variables. Each such variable has just two possible values, the most natural choice being 0 and 1. We illustrate the idea by means of an example, and then make some more general comments about it.

Example 3.12. The following data give the lifetimes in hours of cutting tools, and the lathe speed in rpm at which each tool was used, for each of two makes of cutter A and B (Source: Montgomery and Peck, 1982).

Make A		Make B	
Lifetimes (y_i)	Speeds (x_i)	Lifetimes (y_i)	Speeds (x_i)
18.73	610	30.16	670
14.52	950	27.09	770
17.43	720	25.40	880
14.54	840	26.05	1000
13.44	980	33.49	760
24.39	530	35 62	590
13.34	680	26.07	910
22.71	540	36.78	650
12.68	890	34.95	810
19.32	730	43.67	500

We would like to model the dependence of lifetime on speed, but also if possible take account of any differences that might exist between the two makes of tool. Plotting the data shows that lifetime decreases with lathe speed for both makes, roughly at the same rate for each, but in general the lifetime is greater for make B than make A.

We can take account of the factor 'make' by introducing an indicator variable. Let X_1 be the lathe speed in rpm, and X_2 be the indicator variable for 'make'.

Thus set X_2 equal to 0 for observations on make A and to 1 for observations on make B, and fit the model

$$y_i = \beta_0 + \beta_1 x_{i1} + \beta_2 x_{i2} + \epsilon_i,$$

where the usual assumptions are made about the ϵ_i.

To interpret the parameters of the model, consider each of the two cutter makes in turn. For make A we have $x_{i2} = 0$, so the model is

$$y_i = \beta_0 + \beta_1 x_{i1} + \epsilon_i,$$

i.e. a straight line with intercept β_0 and slope β_1. For make B we have $x_{i2} = 1$, so the model is

$$y_i = \beta_0 + \beta_1 x_{i1} + \beta_2 + \epsilon_i = (\beta_0 + \beta_2) + \beta_1 x_{i1} + \epsilon_i,$$

i.e. a straight line with intercept $(\beta_0 + \beta_2)$ and the same slope β_1 as before.

Thus the postulated model represents two parallel regression lines of cutter lifetime on lathe speed. The parameter β_2 is the *extra* lifetime (on average) experienced by tools of make B, so the two parallel lines are a vertical distance β_2 apart with the line for make B above the one for make A.

Fitting the model in the usual way, we obtain the following results:

Parameter	Estimate	St. Error	t-ratio
β_0	36.986	3.510	10.54
β_1	−0.02661	0.00452	−5.89
β_2	15.004	1.360	11.04

All parameter estimates are thus significant at the 1% level (at least) in the presence of the others, the F-ratio for significance of regression is 76.73, and the coefficient of determination is 0.9003. All indications are thus of an excellent model fit with good predictive properties, and we conclude that tool life decreases at the rate of 2.661 hours per 100 rpm of lathe speed and that cutter make B gives (on average) about 15 hours more lifetime than cutter make A for any fixed speed of operation.

There are two points to be made about such indicator variables in regression.

(i) We are at liberty to choose *any* two distinct values for an indicator variable, not necessarily 0 and 1. The only difference will come in the interpretation of parameters of the model; all other features of the analysis will remain unchanged.

For example, we could set $x_{i2} = -1$ for cutter make A and $x_{i2} = 1$ for cutter make B. The only difference here is that the intercepts of the two parallel regression lines occur at $\beta_0 - \beta_2$ (for make A) and $\beta_0 + \beta_2$ (for make B), but everything else remains unchanged.

(ii) If a qualitative variable has m possible categories ('levels') then we need to set up $m - 1$ indicator variables to allow for it.

For example, if there had been 4 cutter makes A, B, C, D, then define 3 indicator variables X_2, X_3, X_4 having values as follows:

Cutter make	x_{i2}	x_{i3}	x_{i4}
A	0	0	0
B	1	0	0
C	0	1	0
D	0	0	1

With such an assignment, cutter make A forms the 'baseline' reference with regression line having intercept β_0 and slope β_1. Makes B, C and D are then represented by parallel lines that are distances β_2, β_3 and β_4 respectively above the one for A.

3.4 Model building

One of the principal tenets of statistical modelling is that of *parsimony*: we want to use the *simplest* possible model that will achieve the required purpose. In the context of multiple regression, therefore, we have two conflicting objectives:

(i) We want to include as *many* regressors as possible, so as not to 'miss anything', to gain the best predictive power, and to avoid bias.

(ii) We want to include as *few* regressors as possible, so as to end up with the simplest working model, to minimize costs, and to control the variance of predictions.

Inevitably, of course, in practice we have to compromise between these two extremes, so it is necessary to establish a formal framework that will enable us to achieve such a compromise in the best way possible.

There are two common situations encountered in regression studies. The first is where we have fitted a particular model, but suspect that some of the regressor variables in it are redundant. If we have in mind a candidate set of such potentially redundant variables, we might wish to test formally whether the variables in this set 'add anything' to the other variables in the model. Construction of such a test is the focus of the first subsection below. On the other hand, we may have a (potentially large) available set of regressor variables, but no preconceptions: we simply want to find the 'best' set of variables for modelling the response in the given circumstances. This is the situation considered in the remaining two subsections below; the first subsection deals with the case where the number of regressors is not excessive, so that we are able to summarize the results from all possible models that could be formed from the regressors, while the second subsection considers cases where the number of regressors is very large and it is only feasible to fit regressions to a small subset of potential models.

3.4.1 Testing significance of specified subsets of variables

The basic question we want to answer here is: 'Is it worth including the regressors $X_{q+1}, X_{q+2}, \ldots, X_p$ if $X_1, X_2, \ldots, X_q$ are already in the model?' (Note that we can always frame the question in this way, because we can always shuffle the variables round so that the ones we wish to keep in the model are the first q, and the ones whose worth we are questioning come last). We interpret the possible action to be an 'all or nothing' one, i.e. we either keep all the variables $X_{q+1}, X_{q+2}, \ldots, X_p$ in the model or we throw them all out. There are thus two potential models for us to consider: the 'full model' which includes all available regressors,

$$y_i = \beta_0 + \beta_1 x_{i1} + \cdots + \beta_p x_{ip} + \epsilon_i \qquad (3.19)$$

with ϵ_i i.i.d. $N(0, \sigma^2)$, and the 'reduced model' which omits the questionable $X_{q+1}, X_{q+2}, \ldots, X_p$ and includes only $X_1, X_2, \ldots, X_q$,

$$y_i = \beta_0 + \beta_1 x_{i1} + \cdots + \beta_q x_{iq} + \epsilon_i \qquad (3.20)$$

where the same assumptions are made about the ϵ_i.

Formally, the question posed above will be answered by testing the null hypothesis

$$H_0 : \beta_{q+1} = \beta_{q+2} = \cdots = \beta_p = 0$$

against the alternative

$$H_a : \text{at least one } \beta_i \neq 0 \ (i > q)$$

in the full model (3.19). It is thus necesary to obtain a test statistic for these hypotheses.

From intuitive argument, it is evident that if the last $p - q$ regressors *are* important, then the residual sum of squares on fitting the full model (3.19) should be considerably smaller than the residual sum of squares on fitting the reduced model (3.20) (because the latter model assigns systematic variation to the random departure terms, thereby inflating their variance). On the other hand, if the last $p - q$ variables are *not* important, then the two models are 'equivalent' and their respective residual sums of squares should be 'similar'. This intuitive argument suggests that a suitable test statistic for the above null hypothesis is given by the difference in these two residual sums of squares. If we write SS_{EF} for the residual sum of squares of the full model, and SS_{ER} for the residual sum of squares of the reduced model, then SS_{EF} *will always be smaller than* SS_{ER} (since residual sums of squares always decrease on including extra regressors, see Section 3.3.1 above). Thus $E = SS_{ER} - SS_{EF}$ will always be positive, and will provide a suitable test statistic. This difference E is known as the *extra sum of squares* incurred by omitting the subset $X_{q+1}, \ldots, X_p$.

More formally, if we adopt the matrix formulation of models then from (3.10) the full model is

$$y = X\beta + \epsilon, \tag{3.21}$$

where $\beta = (\beta_0, \beta_1, \ldots, \beta_p)'$, X is the $n \times (p+1)$ data matrix whose first column contains ones and whose $(i + 1)$th column contains the n values of X_i, and it is assumed that ϵ has mean vector $\mathbf{0}$ and covariance matrix $\sigma^2 I$. Then if we write $\beta_1 = (\beta_0, \beta_1, \ldots, \beta_q)'$ and denote by X_1 the first $(q + 1)$ columns of X, the reduced model is

$$y = X_1\beta_1 + \epsilon \tag{3.22}$$

where ϵ is assumed to have mean vector $\mathbf{0}$ and covariance matrix $\sigma^2 I$.

Fitting the full model, and referring to the ANOVA table in Section 3.3.4, we see that the full model regression sum of squares is

$$SS_{RF} = \hat{\beta}' X'y - n\bar{y}^2$$

while the full model residual sum of squares is

$$SS_{EF} = y'y - \hat{\beta}' X'y,$$

where y is, as usual, the n-vector of y_i values and $\bar{y}$ is the mean of these values. Correspondingly, on fitting the reduced model we have regression sum of squares

$$SS_{RR} = \hat{\beta}'_1 X'_1 y - n\bar{y}^2$$

and residual sum of squares

$$SS_{ER} = y'y - \hat{\beta}'_1 X'_1 y.$$

Thus the extra sum of squares due to $\beta_2 = (\beta_{q+1}, \ldots, \beta_p)'$ is

$$E = SS_{ER} - SS_{EF} = \hat{\beta}' X'y - \hat{\beta}'_1 X'_1 y. \tag{3.23}$$

Of course, since the regression and residual sums of squares add to S_{yy} for both full and reduced models, it follows that E is equivalently given as the difference $SS_{RF} - SS_{RR}$ of the two *regression* sums of squares. Moreover, since there are p degrees of freedom associated with the regression sum of squares for the full model and q degrees of freedom with the regression sum of squares for the reduced model, there must be $p - q$ associated with the extra sum of squares E.

All these features can be summarized in an expanded ANOVA table, as follows.

Source of Variation	Sum of Squares	Degrees of Freedom	Mean Square
Regression:			
Full model (β)	SS_{RF}	p	$MS_{RF} = \frac{SS_{RF}}{p}$
Reduced model (β_1)	SS_{RR}	q	
Difference $(\beta_2|\beta_1)$	$E = SS_{RF} - SS_{RR}$	$p - q$	$MS_{ext} = \frac{E}{p-q}$
Residual	SS_{EF}	$n - p - 1$	$MS_{EF} = \frac{SS_{EF}}{n-p-1}$
Total	S_{yy}	$n - 1$	

The regression sums of squares SS_{RF}, SS_{RR} and E are often described as being 'due to both β_1 and β_2', 'due to β_1 ignoring β_2', and 'due to β_2 allowing for β_1' respectively. It can be shown that if H_0 is true (i.e. if $\beta_2 = 0$), then E/σ^2 has a chi-squared distribution on $p - q$ degrees of freedom, independently of SS_{EF}/σ^2 (which from earlier sections is known to have a chi-squared distribution on $n - p - 1$ degrees of freedom irrespective of the truth or falsity of H_0). Thus if H_0 is true then the ratio $F_{ext} = MS_{ext}/MS_{EF}$ is consistent with values from an F-distribution on $p - q$ and $n - p - 1$ degrees of freedom, but if H_0 is false then the ratio will be 'bigger' than such values (since, in this case, the numerator E contains extra systematic variation). Hence the statistic F_{ext} provides a test of the extra worth of β_2 over and above β_1 in a similar manner to the overall test of significance of regression based on the ratio $F_{reg} = MS_{RF}/M_{EF}$.

Example 3.13. Consider again the cotton yields and planting days of Example 3.11. We might have wondered whether there really was a point of inflection in the scatter plot, or whether the decrease of yield was effectively linear but with a small 'blip' in the middle of the range. Thus we would want to test the worth of fitting a quadratic and a cubic term on top of a linear one, so our full model would here be the previous cubic

$$y_i = \beta_0 + \beta_1 x_i + \beta_2 x_i^2 + \beta_3 x_i^3 + \epsilon_i$$

for $i = 1, \ldots, 10$, while the reduced model would be the linear

$$y_i = \beta_0 + \beta_1 x_i + \epsilon_i$$

for $i = 1, \ldots, 10$. From the data, we obtain values $S_{yy} = 269.337$ for the total sum of squares, $SS_{EF} = 2.337$ for the residual sum of squares in the full (cubic) model, and $SS_{ER} = 10.926$ for the residual sum of squares in the reduced (linear) model. Here we have $n = 10$, $p = 3$ and $q = 1$ so from this information we can construct the ANOVA table:

Source of Variation	Sum of Squares	Degrees of Freedom	Mean Square	F-Ratio
Regression:				
Full model (cubic)	267.000	3	89.0	228.5
Reduced model (linear)	258.411	1	258.411	
Difference (β_2 and $\beta_3 \mid \beta_1$)	8.589	2	4.295	11.03
Residual	2.337	6	0.3895	
Total	269.337	9		

The two F-ratios are both highly significant, the value 228.5 being tested against an $F_{3,6}$ distribution and showing a highly significant cubic regression, and the value 11.03 being tested against an $F_{2,6}$ distribution and showing a highly significant improvement of the cubic regression over a linear one.

Note that in the above example, there is a natural ordering of the variables which suggests a particular sequence of models: the linear comes first, then the quadratic, then the cubic, and so on. Moreover, it generally makes little sense to have a 'patchwork' of powers; if we decide to include a quartic term, say, then we should include the lower-order linear, quadratic and cubic terms also.

Such a natural ordering of variables is relatively rare, however, and in general we are at liberty to choose whichever variables we like in the two subsets. The important point to remember with the above 'extra sum of squares' test is that the significance of the 'extra' terms depends crucially on the terms already present in the model. If we wish to use the test to decide on the 'absolute' worth of a set of variables, we may need to fit them with and without the presence of the other possible variables. The only situation in which the sequential order of fitting of variables is unimportant is when the regressor variables are *orthogonal* (as defined in Section 3.3.3), which is another benefit of such variables.

Example 3.14. In an investigation to find the factors affecting productivity of plankton in the River Thames, measurements were taken at 17 monthly intervals on the production of oxygen in logarithmic units (Y), the amount of chlorophyll (X_1), and the amount of light (X_2) that was present. The data collected were as follows:

y_i	x_{i1}	x_{i2}	y_i	x_{i1}	x_{i2}
2.16	33.8	329.5	1.16	22.5	358.5
4.13	47.8	306.8	0.61	16.5	210.0
2.84	100.7	374.7	1.94	71.3	361.8
4.65	105.5	432.8	1.70	49.4	300.4
−0.42	33.4	222.9	0.21	19.3	96.9
1.32	27.0	352.1	0.98	71.6	151.8
4.04	46.0	390.8	0.06	13.4	126.0
1.97	139.5	232.6	−0.19	11.8	67.8
1.63	27.0	277.7			

The full model yields the fitted regression equation $y_i = -1.34 + 0.0118x_{i1} + 0.0091x_{i2}$ and associated ANOVA table 1:

Source of Variation	Sum of Squares	Degrees of Freedom	Mean Square	F-Ratio
Regression on X_1 and X_2	24.43	2	12.22	13.79
Residual	12.41	14	0.886	
Total	36.85	16		

Thus fitting both regressors yields a significant regression, but could we do just as well if we used only one of them? The simple regression on X_1 alone yields the fitted equation $y_i = 0.59 + 0.0224x_{i1}$ and ANOVA table 2:

Source of Variation	Sum of Squares	Degrees of Freedom	Mean Square	F-Ratio
Regression on X_1	10.92	1	10.92	6.32
Residual	25.93	15	1.729	
Total	36.85	16		

while the simple regression on X_2 alone yields the fitted equation $y_i = -1.18 + 0.0106x_{i2}$ and ANOVA table 3:

Source of Variation	Sum of Squares	Degrees of Freedom	Mean Square	F-Ratio
Regression on X_2	21.87	1	21.87	21.89
Residual	14.98	15	0.999	
Total	36.85	16		

It thus appears from these ANOVA tables that both X_1 and X_2 provide significant simple regressions, so we might be tempted to infer that both are thus necessary in predicting Y.

This impression is reinforced if we test the extra significance of X_2 when X_1 is already in the model. From ANOVA tables 1 and 2, we obtain 24.43 − 10.92 = 13.51 as the extra sum of squares due to addition of X_2 when X_1 is already in the model. There is 1 degree of freedom for this sum of squares, so $F_{ext} = 13.51 \div 0.886 = 15.25$, which is highly significant against an $F_{1,14}$ distribution.

However, if we consider adding X_1 to a model containing X_2, then ANOVA tables 1 and 3 give an extra sum of squares of 24.43 − 21.87 = 2.56. This value is *not* significant at the 10% level against an $F_{1,14}$ distribution, which indicates that X_1 is *not* necessary if X_2 is already present in the model. Since the simple regression on X_2 is highly significant, with an F-ratio of 21.89 on 1 and 15 degrees of freedom, then we would conclude that the most parsimonious model is given by the one containing X_2 alone.

These sums of squares may be shown in summary form as follows:

Due to	SS
X_1, X_2	24.43
X_1	10.92
$X_2 \vert X_1$	13.51
X_2	21.87
$X_1 \vert X_2$	2.56

3.4.2 Variable selection: examining all subsets

One of the most frequently encountered practical regression problems is that of finding the 'best' model from among a pool of candidate regressor variables. We have already mentioned earlier that this objective involves a compromise between selecting as many regressors as possible (so as not to discard valuable information), and selecting as few as possible (so as to obtain the greatest parsimony). Added to this compromise is the complication that has been exposed in the previous section, namely the fact that any given regressor will have different effects depending upon which variables are present in the model. Thus picking one's way through the various possible models in the manner of Example 3.14 becomes totally impracticable once there are more than about three regressor variables to choose among. Consequently, various automatic procedures have been devised for systematically guiding model choice, and we consider such procedures in this section and the next. We will suppose that there are t regressor variables initially available to the investigator, and that we always have the constant term β_0 in any model that is being examined.

The best approach is obviously one in which we examine all possible regressions. However, with t candidate regressors such an approach involves the examination of 2^t different regressions so brings with it a number of problems. Firstly, even with t relatively small, it will require a great deal of computational effort. Nevertheless, with current computer power and existence of efficient algorithms to generate the required models, it is perfectly feasible to handle up to 30 candidate regressors without too much bother. The second problem, however, is the closely associated one of generating a vast amount of computer output during the procedure. It is clearly unrealistic to produce all the output previously described, for each of 2^t separate regressions. In fact, for the process to be at all viable, we must have no more than a single summary statistic from each regression on which to base our decisions. The third problem is thus the choice of such a statistic. Numerous statistics have been suggested, and in general each statistic is likely to come up with its own 'best' subset of variables. However, three particular statistics have proved to be the most durable ones, and they exhibit a reasonable amount of consistency in their model choices. These are the ones predominantly used in practice, therefore, so we now briefly describe them and their implementation. In each case, the subscript p indicates that there are p terms in the model (i.e. $p - 1$ regressor variables plus the constant term β_0).

(i) **Residual mean square** (RMS_p).

This is just the familiar quantity obtained from the standard ANOVA table, namely the residual sum of squares divided by the residual degrees of freedom. If $p - 1$ regressors are in the model, then we can write the residual sum of squares as $SS_{E(p-1)}$ while the number of residual degrees of freedom is $n - p$, so that $RMS_p = SS_{E(p-1)}/(n-p)$. Typically, RMS_p will initially fall as regressor variables are added to the model (i.e. as p increases), but after a certain point will flatten out and then rise again. ($SS_{E(p-1)}$ always decreases as p increases *but* the reduction on adding a regressor to the model may not compensate for the loss of a degree of freedom when p is moderate or large.) Thus, to use this statistic:

(a) Plot RMS_p against p for each possible model.

(b) Sketch the curve passing through the minimum RMS_p at each p.

(c) Choose the model for which *either* RMS_p is minimum, *or* RMS_p is equal to the residual mean square of the full model (i.e. RMS_{t+1}), *or* RMS_p precedes a sharp upturn in the curve. These three conditions are not necessarily consistent, but in any practical situation at least one will point to a parsimonious choice of model.

(ii) **Coefficient of determination** R_p^2.

Again, a quantity familiar from the standard ANOVA table as the regression sum of squares divided by the total sum of squares (S_{yy}). Thus $R_p^2 = 1 - \frac{SS_{E(p-1)}}{S_{yy}}$. To use this statistic:

(a) Plot R_p^2 against p for each possible model.

(b) Sketch the curve passing through the maximum R_p^2 at each p.

(c) The curve will initially rise steeply, then flatten towards the asymptote at 1. The best model is the one where the most discernible change in slope (i.e. the 'knee') of the plot occurs.

(iii) **Mallow's statistic (C_p).**
This statistic is generally considered to be the most satisfactory of the three, and is defined by

$$C_p = \frac{RSS_p}{s^2} + 2p - n$$

where $s^2 = RMS_{t+1}$, i.e. the estimate of σ^2 from fitting the full model. This statistic is an estimate of the average total error of prediction based on a model containing p terms, and the average total error in turn includes a contribution from prediction bias and one from the precision of the fitted values for the p-term model. When there is negligible bias in fitting a model, C_p is approximately equal to p. To use this criterion, therefore, we plot C_p against p and look for models for which $C_p \leq p$. Of these, the most suitable ones are those with smallest values of C_p.

Example 3.15. An experimental investigation into the heat evolved during the hardening of Portland cement, considered as a function of the chemical composition of the cement, gave the results in Table 3.3 (for further details, see A. Hald *Statistical Theory and Engineering Applications*, 1952, Wiley). The figures indicate the relationship between the heat evolved and the amounts of four chemical compounds contained in the clinkers from which the cement was produced. The compounds were: Tricalcium aluminate (X_1), Tricalcium silicate (X_2), Tetracalcium alumino ferrite (X_3), and Dicalcium silicate (X_4). The heat evolved (Y) after 180 days of curing is given in calories per gram of cement, and the weights of the four components as percentages of the weights of the clinkers. Table 3.4 shows the values of the residual sum of squares $SS_{E(p)}$ after fitting various terms in multiple regression analyses of these data, the degrees of freedom of the residual sums of squares, the residual mean squares RMS_p, the coefficients of determination R^2 and the values of Mallow's statistic C_p.

Figures 3.5, 3.6 and 3.7 show the RMS_p, R_p^2 and C_p plots respectively. From these plots we would conclude that the model with the two regressor variables X_1, X_2 was probably the 'best' one, as it is the model whose RMS_p value is approximately equal to the RMS_{t+1} (full model) value, the model which occurs at the 'knee' of the R_p^2 plot, and also the model which yields minimum C_p value. However, the models including either of the two 3-regressor subsets (X_1, X_2, X_3) or (X_1, X_2, X_4) would also bear further scrutiny, as they are the two models with minimum RMS_p values and the next two in line according to the C_p plot.

Table 3.3 *Data on heat evolved during hardening of cement and on the chemical composition of the cement (Source: Hald, 1952).*

Sample Number (i)	Heat Evolved y_i	% Weight in Clinkers			
		x_{i1}	x_{i2}	x_{i3}	x_{i4}
1	78.5	7	26	6	60
2	74.3	1	29	15	52
3	104.3	11	56	8	20
4	87.6	11	31	8	47
5	95.9	7	52	6	33
6	109.2	11	55	9	22
7	102.7	3	71	17	6
8	72.5	1	31	22	44
9	93.1	2	54	18	22
10	115.9	21	47	4	26
11	83.8	1	40	23	34
12	113.3	11	66	9	12
13	109.4	10	68	8	12

Table 3.4 *Results of all subset regressions for the data in Table 3.3.*

Regressors Included	p	$SS_{E(p)}$	df	RMS_p	R_p^2	C_p
None (β_0 only)	1	2715.76	12	226.31	0.00	442.99
X_1	2	1265.69	11	115.06	0.53	202.55
X_2	2	906.34	11	82.39	0.67	142.49
X_3	2	1939.40	11	176.31	0.29	315.15
X_4	2	883.87	11	80.35	0.67	138.73
X_1, X_2	3	59.90	10	5.99	0.98	2.68
X_1, X_3	3	1227.07	10	122.71	0.55	198.09
X_1, X_4	3	74.76	10	7.48	0.97	5.50
X_2, X_3	3	415.44	10	41.54	0.85	62.44
X_2, X_4	3	868.88	10	86.89	0.68	138.23
X_3, X_4	3	175.74	10	17.57	0.94	22.37
X_1, X_2, X_3	4	48.11	9	5.35	0.98	3.04
X_1, X_3, X_4	4	50.84	9	5.64	0.98	3.50
X_1, X_2, X_4	4	47.97	9	5.33	0.98	3.02
X_2, X_3, X_4	4	73.81	9	8.20	0.97	7.34
X_1, X_2, X_3, X_4	5	47.86	8	5.98	0.98	5.00

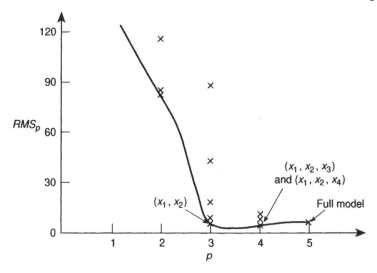

Fig. 3.5 RMS_p *plot derived from Table 3.4.*

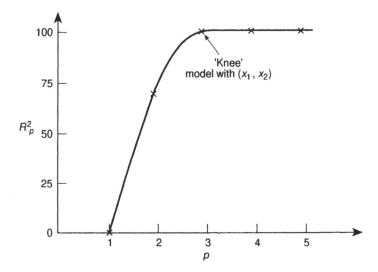

Fig. 3.6 R_p^2 *plot derived from Table 3.4.*

3.4.3 Variable selection: sequential methods

If the number t of candidate regressors is large then it may not be possible to compute all subset regressions, and we need a systematic approach that restricts the number of regressions performed. The most popular methods are *sequential*

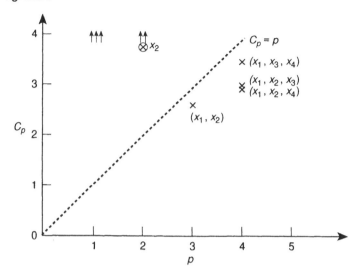

Fig. 3.7 C_p plot derived from Table 3.4.

ones, in which models are either systematically built up by adding variables one by one to the null model comprising just β_0 ('forward selection'), or models are systematically reduced by deleting variables one by one from the full model ('backward elimination'), or a combination of the two processes is employed ('stepwise selection').

If we suppose that at a given stage of the process there are p terms in the model (i.e. β_0 plus a 'base set' of $p-1$ regressor variables) and the residual sum of squares of this model is $SS'_{E(p)}$, then each of the above methods can be described by the following algorithms.

Forward selection

(i) Try each of the remaining $(t - p + 1)$ variables in turn with the existing $(p - 1)$, fit the $t - p + 1$ resulting regressions and note the residual sum of squares value $SS_{E(p+1)}$ in each case.

(ii) Find the variable X_k whose addition to the base set gives the *smallest* value $SS'_{E(p+1)}$ of those obtained in (i), and let RMS'_{p+1} be the residual mean square of the resulting regression.

(iii) Calculate

$$F_0 = \frac{SS'_{E(p)} - SS'_{E(p+1)}}{RMS'_{p+1}}.$$

(iv) If F_0 exceeds a pre-chosen 'F-to-enter' value F_{IN} then add X_k to the base set of regressors and go back to (i) with an increased base set and an increased model containing $p + 1$ terms; otherwise stop the process.

Backward elimination

(i) Try removing each of the $(p-1)$ variables in turn, fit the $p-1$ resulting regressions and note the residual sum of squares value $SS_{E(p-1)}$ in each case.

(ii) Find the variable X_k whose removal from the base set gives the smallest value $SS'_{E(p-1)}$ of those obtained in (i), and let RMS'_{p-1} be the residual mean square of the resulting regression.

(iii) Calculate

$$F_0 = \frac{SS'_{E(p-1)} - SS'_{E(p)}}{RMS'_p}.$$

(iv) If F_0 is less than a pre-chosen 'F-to-remove' value F_{OUT} then remove X_k from the base set of regressors and go back to (i) with a reduced base set and a reduced model containing $p-1$ terms; otherwise stop the process.

Stepwise selection

Proceed as for Forward Selection, but at each stage additionally test to see whether any variable in the current model can be removed. Stop the process when $F_0 < F_{IN}$ for all possible entries *and* $F_0 > F_{OUT}$ for all possible removals.

Comments

We may first note that in both Forward Selection and Backward Elimination the maximum number of regressions to be performed is $t + (t-1) + (t-2) + \cdots + 1 = \frac{1}{2}t(t+1)$, which is considerably fewer than the 2^t of the 'all subsets' approach. Stepwise selection has a slightly greater maximum, but protects against the possibilities that a variable included early in the Forward Selection procedure becomes redundant later on, or that a variable removed early in the Backward Elimination procedure becomes important later on. Backward Elimination may encounter problems in the presence of multicollinearity, as the start of the process requires the fitting of the full model, so the other two approaches are usually preferred.

However, by far the most crucial decision in any of the processes is the choice of the critical values F_{IN}, F_{OUT} (i.e. the choice of the 'stopping rule'). Note first that F_0 in all cases corresponds to the 'extra sum of squares' (Section 3.4.1) F-test for adding or dropping a variable from the current model, so we could use the appropriate percentage point from F-tables for F_{IN} or F_{OUT}. However, the problem with this is that we *select* the variable that has *biggest* F_0 value in each cycle of each process, so F_0 *does not* have an F distribution. One possibility is to choose a conservative value from tables, i.e. one corresponding to a very small significance level. More commonly, a fixed (fairly large) integer value is chosen. For example, the default setting in the MINITAB statistical package is $F_{IN} = F_{OUT} = 4$, but sometimes it is sensible to differentiate the two processes by choosing different values for F_{IN} and F_{OUT} (e.g. if $F_{IN} > F_{OUT}$ then it is relatively harder to add than to remove variables, and vice versa).

Example 3.15 (Cont.). Let us consider application of each of these sequential methods to Hald's cement data. For simplicity let us adopt the default settings $F_{IN} = F_{OUT} = 4$, and note that all relevant quantities in the calculations below can be found in Table 3.4.

First consider Forward Selection. The minimum of the $SS_{E(2)}$ values including just one regressor is 883.87 with X_4, so this is the first candidate for inclusion. Here $F_0 = (2715.76 - 883.87)/80.35 = 22.80$, which clearly exceeds $F_{IN} = 4$ so X_4 is the first member of the base set. Now consider adding each of X_1, X_2, X_3 in turn to X_4. The three resultant regressions have residual sums of squares 74.76, 868.88 and 175.74 respectively. The smallest is 74.76, so we consider adding X_1 to the base set. Here $F_0 = (883.87 - 74.76)/7.48 = 108.17$, also greatly in excess of F_{IN} so X_1 is indeed added. Next we consider adding each of X_2, X_3 in turn to the pair X_1, X_4. The two resulting regressions have residual sums of squares 47.97 and 50.84 respectively, so the variable to be tested is X_2. Here $F_0 = (74.76 - 47.97)/5.33 = 5.02$, again greater than F_{IN} so X_2 is added. Finally we just have the full model to test, for which the residual sum of squares is 47.86 and $F_0 = (47.97 - 47.86)/5.98 = 0.02$. This is clearly negligible, so we don't add X_3 to the base set and finish up with the model that includes X_1, X_2 and X_4.

Turning to Backward Elimination, the first step is to test removal of each variable in turn from the full model. The minimum residual sum of squares among the 3-regressor models is 47.97 (when X_3 is removed) so here $F_0 = (47.97 - 47.86)/5.98 = 0.02$. This is clearly less than $F_{OUT} = 4$, so X_3 is indeed removed from the set to leave X_1, X_2 and X_4. Next we therefore consider the two-regressor sets (X_1, X_2), (X_1, X_4) and (X_2, X_4). The smallest residual sum of squares here is 57.90 (corresponding to the first pair) and $F_0 = (57.90 - 47.97)/5.33 = 1.86$. Again this is less than 4, so we remove X_4 from the current set to leave X_1 and X_2. The smallest residual sum of squares on dropping either of these is 906.34 (dropping X_1) and $F_0 = (906.34 - 57.90)/5.99 = 141.6$. This is clearly much greater than F_{OUT}, so we do not drop X_1. The chosen model thus contains just X_1 and X_2.

Finally, with Stepwise Selection, we find that no deletions are possible at either of the first two stages of the Forward Selection procedure. However, at the third stage we had X_1, X_2 and X_4 in the model. Deleting X_1 gives $F_0 = (868.88 - 47.97)/5.33 = 154.02 > F_{OUT}$, so leave X_1 in. Deleting X_2 gives $F_0 = (74.76 - 47.97)/5.33 = 5.02 > F_{OUT}$, so leave X_2 in. Deleting X_4 gives $F_0 = (59.90 - 47.97)/5.33 = 2.24 < F_{OUT}$, so remove X_4. The chosen model thus contains X_1 and X_2 only.

3.4.4 Comments on variable selection

It is appropriate to conclude this Section with some cautionary comments.

First, as is evident from Example 3.15, the different sequential methods don't necessarily end up with the same final model, and none guarantees that the 'best' subset of regressors of any given size will be chosen. In general, all methods will tend to agree if correlations among regressors are all low. Problems arise as correlations increase, however, and may become acute in the presence of multicollinearity.

In practice, it is better to use these methods to isolate several 'good' models for future examination, rather than accepting what the computer provides as if it were really the 'best' model. One should look at chosen models in relation to what is known about the variables and their context in the practical application, examine the fit of the models critically using the methods of the next section, and discuss the various possibilities with the practitioner who is generating the data before reaching any conclusions.

Finally, one should bear in mind the arbitrariness in the choice of stopping rule, the possibility of bias introduced by model selection, and not ascribe any order of importance to the order in which variables enter or leave the model. The overriding message should be to use subset selection methods with caution!

3.5 Model validation and criticism

As should be evident from earlier discussions, any model is only an approximation to reality; all models involve various assumptions about the data to which they relate; and there may be some portions of the data that are more consonant with the model than others. Thus, whenever a model is fitted to data then all these aspects must be checked critically, especially if the fitted model is to be used for subsequent analysis. In regression studies we should examine the residuals to check on model assumptions and to identify any *outliers* that might be present, we should inspect the regressor variables for evidence of any *influential observations*, and if there are replicate observations at particular combinations of regressor variable values then we can formally test for *lack of fit* of the regression. We now briefly consider each of these aspects, and indicate some transformations that can be used to improve the fitted relationship in cases where assumptions are violated.

3.5.1 Examination of residuals

We here return to the basic regression model as given in (3.10), i.e. $y = X\beta + \epsilon$ where ϵ has mean vector $\mathbf{0}$ and covariance matrix $\sigma^2 I$. In Section 3.3.4, we defined the vector of *fitted values* $\hat{y} = (\hat{y}_1, \ldots, \hat{y}_n)' = X\hat{\beta}$ and the vector of *residuals* $e = (e_1, \ldots, e_n)' = y - \hat{y}$. We also showed in equations (3.13) and (3.14) respectively that $\hat{y} = Hy$ and $e = (I - H)y$, where $H = X(X'X)^{-1}X'$ is known as the 'hat' matrix.

Applying properties (ii) and (iii) of Section 3.3.2 to this definition of e, and simplifying the algebra, we find that $E(e) = 0$ and $\text{cov}(e) = \sigma^2(I - H)$ if the assumptions of model (3.10) are true. Thus if we write the (i, j)th element of H as h_{ij}, it follows that

$$\text{var}(e_i) = (1 - h_{ii})\sigma^2 \tag{3.24}$$

and

$$\text{cov}(e_i, e_j) = -h_{ij}\sigma^2. \tag{3.25}$$

Intuitively, looking at the residuals should give an indication of the quality of fit of the regression: if all the residuals are 'small' then the fit is good, but if some or all are 'large' then the fit leaves something to be desired. Unfortunately the raw residuals are not easy to interpret directly, as their scale of values depends on the range and scales of the response and regressor variables. An 'average' variance of the residuals is given by the residual mean square, s^2, which estimates the unknown departure variance σ^2, so a simple scaling of the residuals is to divide each by the average standard deviation s. However, equation (3.24) shows that the residuals do not have constant variance. In view of this, it is preferable to work with the *standardized* residuals (sometimes referred to as *studentized* residuals) defined by

$$e_i' = \frac{e_i}{s\sqrt{(1 - h_{ii})}}. \tag{3.26}$$

Furthermore, equation (3.25) shows that the residuals are not independent. However, the covariance between any two of them will usually be low, particularly if the sample size n from which the model has been fitted is large. Finally, if we assume normality of the ϵ_i in (3.10) then the fact that the residuals are linear functions of the y_i implies normality of the e_i' also. Thus, *if all the assumptions underlying the regression model are correct*, we can view the e_i' as iid $N(0, 1)$ random variables. Graphical inspection of the e_i' will quickly show up any violation of model assumptions or any model inadequacy, as such situations will induce deviation from iid $N(0, 1)$ behaviour on the part of the standardized residuals. Some useful procedures are as follows.

Model assumptions

- To check on the assumption of normality, plot the ranked standardized residuals against inverse normal cumulative distribution values as described for the simple regression case in Section 3.2.6 (departure from normality again being indicated by deviation of the plot from a straight line), or check on the proportion of the sample lying inside the ranges ± 2.5, ± 2.0 etc (which should be approximately 99%, 95% etc under normality).

- To check on the assumption of constant variance, plot standardized residuals e'_i against their corresponding fitted values $(\hat{y}_i)$. The points should appear randomly and evenly scattered about zero if the assumption is justified, but will exhibit systematic changes in spread otherwise.
- To check on the assumption of independence, plot the standardized residuals against the serial order in which the observations were taken. Again, random scatter of points indicates that the assumption is valid but any visible 'trend' suggests that it is violated

Model adequacy

Here we plot the e'_i against individual regressor variables. Adequacy of model is indicated if all plots exhibit random scatter of equal 'width' about zero, but model inadequacy is shown up if any plots exhibit some systematic pattern. Non-linearity when residuals are plotted against any regressors in the model suggests that higher-order terms involving those regressors should be added to the model. Systematic patterns exhibited when residuals are plotted against regressors that are *not* in the model suggests that those regressors should be added to the model.

Outliers

Outliers are observations that are not well fitted by the assumed model. Such observations will have large residuals. A crude 'rule of thumb' is that an observation with a standardized residual greater than 2.5 in absolute value indicates a possible outlier, and the source of the data should be investigated. If there is doubt about the accuracy or veracity of the observation, then it should be omitted and the model should be refitted. If a large difference in the model parameter estimates is thereby obtained, then the observation is said to be *influential*. Influential outliers are more important than ones which do not cause much perturbation in model parameters on omission, because they raise the possibility of model instability. However, it is not only outliers that can be influential, so this topic deserves further discussion.

3.5.2   Influential observations

An influential observation is defined to be one which, for whatever reason, causes a large change in some or all of the estimated regression parameters when it is omitted from the data set. We have seen above that sometimes a large outlier is an influential observation. However, outliers reflect disparities between observed and fitted response variable values, and these are not the sole causes of model instability. If an observation is well separated from the others in terms of its values on the regressor variables, then this observation is also likely to influence the fitted regression model. Finally, a combination of regressor variable separation and response variable disparity also has the potential for influencing regression fit. If a small subset of data points is likely to exert a large influence on the parameter estimates and predictions of a fitted model, we want some simple measures to enable us to locate such points and assess their influence.

Observations well separated from the others in terms of their regressor variable values will have a large value of h_{ii}. We call h_{ii} the *leverage* of the ith observation. As a rough guide, observations with $h_{ii} > 3(p + 1)/n$ are influential, where p is the number of regressor variables in the model and n is the number of observations.

However, the drawback with leverage as a measure of influence is that it only takes into account the data configuration regarding the *regressor* variables, so it may miss those observations whose influence arises partly or wholly through their response variable values. Various other measures of influence have therefore been proposed, the most popular of which is probably Cook's distance

$$D_k = \frac{1}{(p+1)s^2} \sum_{i=1}^{n} (\hat{y}_{i(k)} - \hat{y}_i)^2, \qquad (3.27)$$

where $\hat{y}_{i(k)}$ is the fitted value for the ith observation when the kth observation is omitted from the model-fitting process. This statistic is a scaled measure of the difference between the fitted values with and without the kth observation included in the fitting process, so a large value of D_k indicates that the kth observation is influential. Also, it can be shown that

$$D_k = e_k'^2 h_{kk}/([1 + p][1 - h_{kk}]),$$

so the statistic can be computed easily for all k without having to redo any regression fits.

3.5.3 Replicate observations and lack of fit

Suppose we have a simple regression in which a response variable is being modelled by a linear function of a single regressor variable, so that we have a situation such as that represented by equation (3.1), and we have fitted the model and constructed the ANOVA table from it. Now suppose that we should in fact have fitted a quadratic function of the regressor variable. Then there is a systematic *lack of fit* between the straight line and the quadratic curve, and the ANOVA residual sum of squares will be a mixture of two contributions: the deviations between the line and the curve, plus the departures of the observations from the (true) curve. The departure terms (i.e. the ϵ_i) in the fitted model have subsumed the quadratic term of the true model, so the ANOVA residual mean square will consequently be inflated and will overestimate σ^2.

However, if there are replicate Y observations at some of the values of the regressor variable X, then we can always obtain a *correct* estimate of σ^2, known as the *pure error* term. The ANOVA residual sum of squares can then be partitioned into a 'pure error' component plus a remainder, known as the 'lack of fit' component and the latter can be tested for significance. The details are as follows.

Suppose there are k values $x_1, x_2, \ldots, x_k$ of X at each of which Y is replicated, with n_i replicate values of Y at x_i. Let y_{ij} denote the jth value of Y at x_i ($j = 1, \ldots, n_i; \; i = 1, \ldots, k$), and write $\bar{y}_i = \frac{1}{n_i}\sum_j y_{ij}$ for the mean of the Y values at x_i. Then the 'pure error' sum of squares is the sum of squared deviations *within replicates*, i.e.

$$SS_e = \sum_{i=1}^{k}\sum_{j=1}^{n_i}(y_{ij} - \bar{y}_i)^2,$$

and this sum of squares has $d_e = \sum_i(n_i - 1)$ degrees of freedom. Thus if SS_E denotes, as usual, the residual sum of squares for the regression ANOVA (with $d_E = n - 2$ degrees of freedom), then the difference $SS_E - SS_e$ is the 'lack of fit' sum of squares and it has $d_E - d_e$ degrees of freedom. The correct estimate of σ^2 is *always* given by the pure error mean square SS_e/d_e, so we test the lack of fit by calculating the ratio of the lack of fit mean square to the pure error mean square, i.e. $F_0 = \frac{(SS_E - SS_e)\div(d_E - d_e)}{SS_e\div d_e}$, and referring the calculated value to the $F_{d_E - d_e, d_e}$ distribution for significance.

The same general principle holds for testing lack of fit in multiple regression. However, in this case we need replicate Y observations at more than one *combination* of regressor variables, so the situation is less common in practice and we do not give details of the computations.

Example 3.16. Soap manufacturers are interested in the relationship between the amount of suds produced and the weight of soap dissolved. An experiment was conducted in which the weight (gm) of soap (X) was varied, and the height (cm) of suds (Y) was measured in a standard dish subjected to a given amount of agitation. The following data were obtained:

x_i	1.3	2.0	2.7	3.3	3.7	4.0	4.7	5.0	5.3	5.7	6.0	6.3	6.7
y_{ij}	1.8	1.5	2.2	1.8	1.7	2.2	1.9	1.8	2.1	3.4	3.0	3.0	5.9
	2.3	2.8		3.8	3.7	2.8	3.2		2.8		3.2		
						2.8	5.4		3.5				

A simple linear regression yielded the fitted relationship $\hat{y} = 1.436 + 0.338x$, with associated ANOVA table

Source	S.S.	d.f.	M.S.	Ratio
Regression	6.326	1	6.326	6.569
Residual	21.192	22	0.963	
Total	27.518	23		

The straight line regression is thus significant at the 5% level (the critical value of $F_{1,22}$ distribution being 4.30), but is it a reasonable equation to fit? Since most x_i values have 2 or 3 replicate Y observations, we can split the residual sum of squares into lack of fit and pure error components. We find the following quantities making up the pure error component:

x_i	1.3	2.0	3.3	3.7	4.0	4.7	5.3	6.0	Total
$\sum(y_{ij} - \bar{y}_i)^2$	0.125	0.845	2.0	2.0	0.24	6.26	0.98	0.02	12.47
df	1	1	1	1	2	2	2	1	11

(For example, the three repeat values at $x_i = 4.7$ have mean 3.5 and hence contribution $(1.9 - 3.5)^2 + (3.2 - 3.5)^2 + (5.4 - 3.5)^2 = 6.26$.) Thus we have $SS_e = 12.470$ and $d_e = 11$, so the lack of fit component is $21.192 - 12.470 = 8.722$, with $22 - 11 = 11$ degrees of freedom. The ANOVA table can therefore be rewritten

Source	S.S.	d.f.	M.S.	Ratio
Regression	6.326	1	6.326	
Lack of fit	8.722	11	0.793	0.699
Pure error	12.470	11	1.134	
Total	27.518	23		

The lack of fit F-ratio is $0.793 \div 1.134 = 0.699$, which is not significant when referred to the $F_{11,11}$ distribution. Thus there is no reason to doubt the adequacy of the straight-line model, and we can use the original residual mean square of 0.963 as an estimate of σ^2.

3.5.4 Remedial actions

In this section, we have considered a number of checks and diagnostics that should be carried out as a matter of course in any regression analysis. If all the calculations suggest that the model is adequate and that any assumptions made in fitting it are reasonable, then fine. But what do we do if any of these calculations highlight a problem with the model or a failure of assumptions?

Sometimes the necessary remedial action will be obvious. For example, if residual plots suggest the necessity of modifying the model in some way (e.g. including extra regressors, or higher-order terms of some existing ones) then we should try fitting the modified model and conduct the same checks and residual plots on the new fit. Similarly, if there is a significant lack of fit in a situation with replicate observations then a residual plot should suggest a better model to try. The whole procedure is then iterated and the model refined until, eventually, an acceptable fit is achieved.

However, what can we do in cases (such as failure of assumptions) where no obvious remedy is suggested by the diagnostic procedure? In such cases a possible way forward is to *transform* one or more of the variables, conduct a new regression on the transformed variables, and then if necessary back-transform any results obtained from this regression to express them in terms of the original variables. The most common failures of assumptions are with either non-normality or non-constancy of variance of Y. In fact, these two violations often go together: if Y is non-normal, then it usually comes from a distribution whose variance is functionally related to its mean. Transforming Y to some new variable Z which has approximately constant variance will often also make Z approximately normal, and we can then regress Z on the X variables in the standard way. Some common transformations exploit either the type of data or the mean–variance relationship. The most useful ones in general statistical applications are:

(i) the arcsine transformation, $z = \sin^{-1}(\sqrt{y})$, appropriate either if the responses are proportions with constant denominators or if $\sigma^2 \propto \mu(1 - \mu)$;

(ii) the square root transformation, $z = \sqrt{y}$, appropriate either if the responses are counts or if $\sigma^2 \propto \mu$; and

(iii) the log transformation, $z = \log y$, appropriate if $\sigma^2 \propto \mu^2$.

However, the distributional assumption is usually paramount, in which case we are better served by fitting a *generalized* linear model. Such models form the subject matter of Chapters 5–7.

3.6 Comparison of regressions

Situations often arise in which a response variable Y and a set of regressor variables $X_1, X_2, \ldots, X_p$ are measured in a set of k distinct *a priori* groups, and we wish to examine to what extent the relationship between Y and the X_i differs between the groups. Once again, as with the case of replicate observations in Section 3.5.3 above, we will specify the details for the case of a single regressor variable X only; extension to the multiple regression case follows in the obvious way.

To fix ideas, consider a specific example. We have available the following data on photosynthesis rate (Y) and radiation (X) for three levels (Low, Medium and High) of water availability, and we wish to determine how similar the relationship between photosynthesis and radiation is at the three water levels.

Water Level 1 (Low)		Water Level 2 (Medium)		Water Level 3 High	
y_i	x_i	y_i	x_i	y_i	x_i
210	2.5	210	5.2	247	6.8
353	4.8	352	6.6	350	9.8
448	7.5	440	10.1	346	13.8
590	5.8	621	12.2	450	15.3
555	9.6	695	14.8	560	19.8

Clearly, if the relationships are completely different then we need to fit separate regression lines to the five (x_i, y_i) pairs at each water level, while if the relationships are exactly the same then we can fit a single regression line to all fifteen pairs. What intermediate possibilities exist? One is for the intercepts of the three regression lines to coincide but the slopes of the three lines to differ, while another is for the three intercepts to differ but the slopes to be the same. The former is not a very common option, as it requires a very specific type of data for it to be realistic. However, the latter (i.e. parallel regression lines) is a very popular intermediate step, as it implies that the relationship between Y and X is the same in each group, but the 'base-line' value differs between groups.

In general, let us suppose that there are k groups of data, with n_i pairs of values in the ith group $(i = 1, \ldots, k)$ and $N = \sum_i n_i$. To investigate the similarities in the relationships between groups we carry out the following steps.

(i) Fit a single regression line to all the data. Let the regression sum of squares be denoted by SS_R, the total sum of squares by S_{yy}, and the residual sum of squares by $SS_{E(1)}$. The numbers of degrees of freedom corresponding to these three sums of squares are thus 1, $N-1$ and $N-2$ respectively.

(ii) Fit parallel lines in the k groups. This is most easily done by defining $k-1$ dummy variables $Z_1, Z_2, \ldots, Z_{k-1}$ to denote an individual's group membership, as explained in Section 3.3.8, and then fitting the regression model

$$y_i = \beta_0 + \beta_1 z_{i1} + \cdots + \beta_{k-1} z_{i,k-1} + \beta x_i + \epsilon_i.$$

Having fitted this model, $\hat{\beta}$ gives the common slope of the regression line in each group, $\hat{\beta}_0$ gives the intercept of the line for group 1, and $\hat{\beta}_0 + \hat{\beta}_j$ gives the intercept of the line for group $j+1$ $(j = 1, \ldots, k-1)$. Denote the residual sum of squares for this model by $SS_{E(2)}$. This has $N-k-1$ degrees of freedom (since the model has k regressor variables), and the difference $SS_p = SS_{E(1)} - SS_{E(2)}$ is the 'extra sum of squares due to parallel lines given a single line'. This extra sum of squares has $(N-2) - (N-k-1) = k-1$ degrees of freedom, and measures the improvement in fit achieved when parallel lines replace the single line.

(iii) Fit a separate line in each group, and add up the residual sums of squares from the separate regressions to give the value $SS_{E(3)}$. This has $\sum_i (n_i - 2) = N - 2k$ degrees of freedom, and the difference $SS_s = SS_{E(2)} - SS_{E(3)}$ is the 'extra sum of squares due to separate lines given parallel ones'. This extra sum of squares has $(N-k-1) - (N-2k) = k-1$ degrees of freedom, and measures the improvement in fit achieved when the constraint of parallelism of lines is relaxed to allow arbitrary slopes in each group.

The above calculations are then summarized in the ANOVA table:

Source of Variation	Sum of Squares	Degrees of Freedom	Mean Square	Ratio
Single line	SS_R	1	$MS_R = SS_R$	$F_1 = \frac{MS_R}{MS_E}$
Parallel given single	SS_p	$k-1$	$MS_p = \frac{SS_p}{k-1}$	$F_p = \frac{MS_p}{MS_E}$
Separate given parallel	SS_s	$k-1$	$MS_s = \frac{SS_s}{k-1}$	$F_s = \frac{MS_s}{MS_E}$
Residual	$SS_{E(3)}$	$N-2k$	$MS_E = \frac{SS_{E(3)}}{N-2k}$	
Total	S_{yy}	$N-1$		

We test the ratio in each line of this ANOVA against an F-distribution in the usual way, using the degrees of freedom in that line and in the residual line for the F degrees of freedom. A significant F_1 indicates that there is an overall relationship between Y and X, a significant F_p indicates that parallel lines describe the data better than a single line, and a significant F_s indicates that separate lines describe the data better than parallel lines. In the interests of parsimony we seek the simplest model to describe the data, so stop at the first stage beyond which there are no significant F-ratios.

Example 3.17. For the photosynthesis and radiation data given above, we find as follows.

(i) Fitting a single line to all the data, $S_{yy} = 321378$, $SS_R = 111845$ and $SS_{E(1)} = 209533$.
(ii) Fitting 3 parallel lines, $SS_{E(2)} = 78921$, so $SS_{E(1)} - SS_{E(2)} = 209533 - 78921 = 130612$.
(iii) Fitting 3 separate lines, $SS_{E(3)} = 48106$, so $SS_{E(2)} - SS_{E(3)} = 30815$.

Here we have $k = 3$ and $N = 15$, so the ANOVA table is

Source of Variation	Sum of Squares	Degrees of Freedom	Mean Square	Ratio
Single line	111845	1	111845	$F_1 = 20.93$
Parallel lines given single	130612	2	65306	$F_p = 12.22$
Separate lines given parallel	30815	2	15408	$F_s = 2.88$
Residual	48106	9	5345	
Total	321378	14		

We find both F_1 and F_p to be significant at the 1% level, so there is strong evidence that photosynthesis rate increases with increasing radiation level, and there is a significant improvement in using the parallel lines model over the single line model. However, the value 2.88 is not significant when compared with an $F_{2,9}$ distribution, so there is no significant improvement in fitting separate lines over parallel lines. The only difference between the water levels is thus in mean photosynthesis rather than in its relationship with radiation.

The parallel lines regression in terms of the dummy variables has the fitted equation

$$\hat{y} = 223 - 96.8z_1 - 284z_2 + 34.5x,$$

so disaggregating into the three parallel lines by substituting the values $(0, 0)$, $(1, 0)$ and $(0, 1)$ for (z_1, z_2) respectively in this fitted equation, we obtain the equations

$$y = 223 + 34.5x \text{ for water level 1 (low)},$$

$$y = 126.2 + 34.5x \text{ for water level 2 (medium)},$$

and

$$y = -61 + 34.5x \text{ for water level 1 (high)}.$$

Confidence intervals for parameters, and for differences of parameters between groups, can be found in the usual way.

Multiple regressor variables

We can conduct a similar comparison of regressions when there is more than one regressor variable, but there are now many more possibilities open to us (e.g. equal regression coefficients across groups for some variables but not for others). There are thus many possible subdivisions of the ANOVA sum of squares, and the precise models chosen to test will depend on the context of the data. We therefore do not give any details, but merely note that the general principles are exactly the same as above.

3.7 Non-linear models

As has already been stressed earlier, non-linear models are those that are non-linear *in their parameters*. Typical examples are:

(i) Mitscherlich's equation for general asymptotic relationships

$$y_i = \alpha - \beta e^{-\gamma x_i} + \epsilon_i \qquad (\gamma > 0);$$

(ii) the logistic growth curve for sigmoidal relationships

$$y_i = \frac{\gamma \exp(\alpha + \beta x_i)}{1 + \exp(\alpha + \beta x_i)} + \epsilon_i.$$

Sometimes a non-linear model is *linearizable* by transformation, but this often depends crucially on the departure term. For example, the model which has systematic component $\exp(\alpha + \beta x)$ is linearizable when departures are assumed to be *multiplicative*, because the model is then $y_i = \epsilon_i \exp(\alpha + \beta x_i)$, which is equivalent to the linear model $\log y_i = \alpha + \beta x_i + \delta_i$, where $\delta_i = \log \epsilon_i$. However, if the departures are assumed to be additive then the model is $y_i = \exp(\alpha + \beta x_i) + \epsilon_i$, which cannot be reduced to linear form by transformation.

If a model is linearizable, then the linear version can be fitted by the methods described earlier in this chapter. However, residuals of the fitted model should always be checked for acceptability and if there are any problems then non-linear methods should be tried.

If the model is not linearizable, then it must be fitted by *non-linear least squares*. In general, suppose the model to be fitted is

$$y_i = f(x_{i1}, x_{i2}, \ldots, x_{ip}; \theta_1, \ldots, \theta_k) + \epsilon_i$$

where $f(\cdot)$ is the non-linear function of the regressor variables $X_1, \ldots, X_p$ and unknown parameters $\theta_1, \ldots, \theta_k$, and the ϵ_i are assumed, as usual, to be iid $(0, \sigma^2)$ random variables. For notational simplicity, let us write x_i to denote the set of values $(x_{i1}, \ldots, x_{ip})$, and $\boldsymbol{\theta}$ to denote the set $(\theta_1, \ldots, \theta_k)$. Then the model is

$$y_i = f(x_i; \boldsymbol{\theta}) + \epsilon_i,$$

the sum of squared departures is

$$S = \sum_{i=1}^{n} \{y_i - f(x_i; \boldsymbol{\theta})\}^2,$$

and the least-squares estimator of $\boldsymbol{\theta}$ is the value $\boldsymbol{\theta}$ minimizing S. One way of finding such an estimator directly is to use a standard computer minimization routine such as steepest descent, Gauss–Newton or Marquart's method to minimize S numerically with respect to $\boldsymbol{\theta}$. Alternatively, differentiating S with respect to each of the θ_i in turn and setting the resulting expressions to zero yields the set of simultaneous equations

$$\sum_{i=1}^{n} \left[\{y_i - f(x_i; \boldsymbol{\theta})\} \left(\frac{\partial f}{\partial \theta_j} \right)_{\boldsymbol{\theta} = \hat{\boldsymbol{\theta}}} \right] = 0 \qquad (j = 1, \ldots, k).$$

In general, an iterative numerical method is needed to solve these equations.

Example 3.18. Consider fitting the simplified Mitscherlich equation

$$y_i = \alpha(1 - e^{-\gamma x_i}) + \epsilon_i$$

to a set of n (x_i, y_i) pairs. Here we have

$$S = \sum_{i=1}^{n} \{y_i - \alpha(1 - e^{-\gamma x_i})\}^2,$$

so that

$$\frac{\partial S}{\partial \alpha} = -2 \sum_{i=1}^{n} [\{y_i - \alpha(1 - e^{-\gamma x_i})\}(1 - e^{-\gamma x_i})]$$

and

$$\frac{\partial S}{\partial \gamma} = -2 \sum_{i=1}^{n} [\{y_i - \alpha(1 - e^{-\gamma x_i})\}(\alpha\gamma)e^{-\gamma x_i}].$$

The normal equations are thus

$$\sum_{i=1}^{n} [\{y_i - \hat{\alpha}(1 - e^{-\hat{\gamma} x_i})\}(1 - e^{-\hat{\gamma} x_i})] = 0 \tag{3.28}$$

and

$$\sum_{i=1}^{n} [\{y_i - \hat{\alpha}(1 - e^{-\hat{\gamma} x_i})\}(\hat{\alpha}\hat{\gamma})e^{-\hat{\gamma} x_i}] = 0. \tag{3.29}$$

Moreover, (3.28) leads to

$$\hat{\alpha} = \frac{\sum y_i(1 - e^{-\hat{\gamma} x_i})}{\sum (1 - e^{-\hat{\gamma} x_i})^2}. \tag{3.30}$$

These equations suggest two alternative schemes in which successive pairs $(\hat{\alpha}_i, \hat{\gamma}_i)$ are generated iteratively until convergence. The first scheme has the following steps.

(i) Obtain initial estimate $\hat{\gamma}_0$.
(ii) Given $\hat{\gamma}_i$, find $\hat{\alpha}_i$ from (3.30).
(iii) Substitute $\hat{\alpha}_i$ into (3.29) and solve for $\hat{\gamma}_{i+1}$.
(iv) Cycle steps (ii) and (iii) until $\hat{\alpha}_i$, $\hat{\gamma}_i$ pairs converge.

The alternative scheme has the following steps.

(i) Obtain initial estimate $\hat{\alpha}_0$.
(ii) Given $\hat{\alpha}_i$, find $\hat{\gamma}_i$ from (3.29).
(iii) Substitute $\hat{\gamma}_i$ into (3.30) to obtain $\hat{\alpha}_{i+1}$.
(iv) Cycle steps (ii) and (iii) until $\hat{\alpha}_i$, $\hat{\gamma}_i$ pairs converge.

We can make a number of general observations about non-linear models and methods of fitting them.

(a) Initial values for an iterative technique such as in Example 3.18 can be found in a number of ways. For example,

  (i) use knowledge of the parameters to estimate initial values from a plot of the data;
  (ii) fit a regression to the nearest 'linear' approximation of the model and use the resulting parameter estimates as initial values;

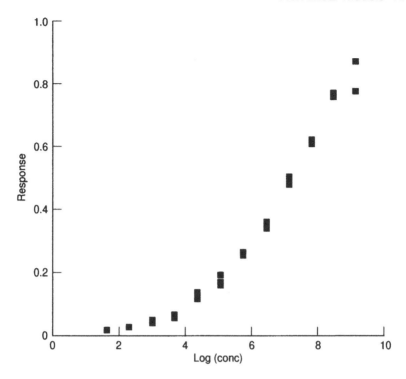

Fig. 3.8 *Plot of absorbance against the logarithm of concentration in a radioim-munoassay experiment.*

      (iii) solve the model equation at k 'typical' (x_i, y_i) points to give the initial values.

(b) In non-linear models, the parameter estimates will not in general be linear combinations of the observations y_i, so even if we assume normality of the y_i the $\hat{\theta}_i$ will *not* be normally distributed. Also, the covariance matrix of θ will no longer be $(X'X)^{-1}\sigma^2$.

(c) We can still find the total sum of squares and the residual sum of squares as usual, and hence build up an ANOVA table, but the above non-normality implies that we cannot use F-tests to test significance in this ANOVA. However, we *can* use the residual mean square as an estimator of σ^2; it is no longer unbiased, but it does give a good approximation to σ^2.

(d) Finally, if we assume normality of the y_i then the least squares estimators are also maximum likelihood ones, so we can obtain approximate standard errors of the $\hat{\theta}_i$ from asymptotic properties of maximum likelihood estimators.

Example 3.19. To illustrate some of these ideas, consider the data given below.

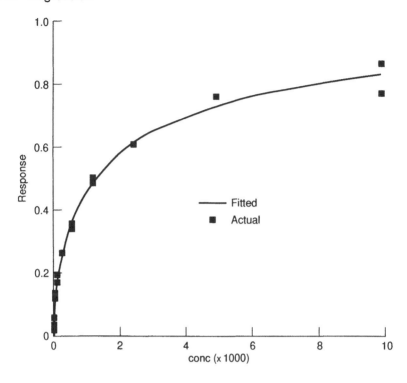

Fig. 3.9 *Plot of actual values and fitted model for the radioimmunoassay data.*

These data were obtained from a radioimmunoassay experiment in which a standard solution was diluted to a number of concentrations, and the absorbance was measured at each concentration. The absorbance is the response variable Y and the concentration is the explanatory variable X.

x_i	y_i	x_i	y_i
10000	0.880	156	0.192
10000	0.784	156	0.173
5000	0.776	78	0.125
5000	0.769	78	0.138
2500	0.622	39	0.070
2500	0.614	39	0.064
1250	0.500	20	0.050
1250	0.488	20	0.044
625	0.347	10	0.029
625	0.356	10	0.029
312	0.263	5	0.018
312	0.260	5	0.018

A plot of y_i against $\log x_i$ yields the sigmoid shape in Figure 3.8. A model used widely in radioimmunossay work is the four-parameter logistic model given by

$$y_i = \alpha + \frac{\beta - \alpha}{1 + (x_i/\gamma)^{-\delta}} + \epsilon_i \qquad (i = 1, \ldots, n) \qquad (3.31)$$

where α, β, γ, δ are the four parameters and ϵ_i is the ith departure term. The departures are as usual assumed to be iid $N(0, \sigma^2)$.

Plotting the systematic part of this model on the $\log x$ scale shows that α is the intercept, β is the asymptote, and $y = \frac{1}{2}(\alpha + \beta)$ when $x = \gamma$. Matching these values to their counterparts in Figure 3.8 gives initial estimates 0.0, 0.9 and 1096 for α, β and γ respectively. To obtain an initial estimate of δ, we note that the systematic part of the model can be rewritten as

$$\log\left(\frac{y - \alpha}{\beta - y}\right) = -\delta \log \gamma + \delta \log x.$$

Hence if we plot $\log(\frac{y-\alpha}{\beta-y})$ against $\log x$ using the initial values of α and β, then the slope of the resulting approximate straight line gives an initial estimate of δ. Using this method in the present case yields an initial estimate of 0.7.

A numerical non-linear least squares program then produced the following parameter estimates and approximate standard errors (from the matrix of second derivatives):

parameter	estimate	st. error
α	0.0025	0.0138
β	1.0736	0.0697
γ	1577.9	309.87
δ	0.711	0.0645

The closeness of the estimate of alpha to zero is intuitively justifiable: no response at zero concentration. The residual mean square in the ANOVA table estimates σ^2 as 0.000414. A plot of actual values and fitted model (on the x rather than $\log x$ scale) is shown in Figure 3.9; no untoward points were found on inspecting residuals, and there were no convergence problems with the numerical algorithm. The model proved to be a successful fit to the data.

4

Normal response and qualitative explanatory variables: analysis of variance

4.1 Motivation

In the previous chapter we considered situations in which a quantitative response variable Y is modelled as a linear function of one or more quantitative explanatory variables $X_1, \ldots, X_p$ plus a departure term, the aim usually being either to explain the mechanism generating Y values or to predict future Y values given the corresponding values of the explanatory variables.

In this chapter we turn attention to situations that are equally common in practice, but in which the emphasis of the analysis is somewhat different. These situations are ones in which observations are again taken on a quantitative response variable Y, but quantitative explanatory variables X_i are no longer available. Instead, the observations on Y are taken either under a variety of external conditions or in distinct *a priori* groups, and the primary aim of the analysis is to determine whether these external conditions or *a priori* groups have any systematic effects on the resulting values of Y. If it can be established that there *are* systematic effects caused by groups or conditions, then a secondary aim of the analysis is to examine the nature of these effects and report on any implications arising from them.

The following specific examples will help to fix ideas.

- Subjects volunteering for an experiment in a psychology laboratory are first divided into four groups according to their age. They are all then set the same task to do, and the time each takes to complete the task is measured in minutes. Is there a systematic difference between age groups in the times taken? If, additionally, each individual is categorized as male or female, is there any systematic gender difference in completion times? If there is a gender difference, is it independent of the subject's age, or does it vary differentially across the age groups?

- Six different varieties of corn are planted, and after a given time samples of each variety are harvested and weighed. Is there a systematic difference between the weights of different varieties? If each variety has been planted in three different types of soil on each of eight farms, do the weights differ between farms and types of soil as well as between varieties?

- Five panellists at a Food Research Institute are asked to rate each of six meat samples for 'succulence' on a scale from 0 to 10, the latter expressing the 'most succulent' extreme. Is there a systematic difference between panellists in respect of the scores they allocate to the samples? Or between the samples in terms of their 'succulence'?

In all such cases there are *qualitative* or *categorical* explanatory variables instead of quantitative ones. Thus 'age group' with four categories and 'gender' with two categories are the two such explanatory variables in the first example; 'varieties' with six categories, 'soil types' with three categories, and 'farms' with eight categories are the three explanatory variables in the second example; and 'panellists' with five categories and 'meat samples' with six categories are the explanatory variables in the last example.

Since we have seen in Section 3.3.8 that categorical variables can be expressed as dummy regressor variables, it follows that all such examples can actually be reformulated as multiple regression problems and the methods of Chapter 3 can be applied to effect the analysis. We will come back to this aspect later in this chapter. To develop methods of analysis, however, we will eschew the regression formulation and instead concentrate on an approach based on the partition of sums of squares among the different external conditions/groups. This approach gives better insight into the process underlying the analysis, and in many cases leads to relatively simple calculations that can be done on a pocket calculator (whereas the corresponding regression calculations are much more opaque and always require the use of a computer). For complex arrangements, however, we have no option but to fall back on the regression approach so the details are included at the end of the chapter.

On a historical note, many of the techniques that follow were first developed by R. A. Fisher in the early part of the 20th century specifically in the context of agricultural experimentation. In view of the focus on experimentation, it is common to encounter in text books the term 'treatments' for what we have referred to as 'groups' or 'conditions' above. Moreover, since most agricultural experiments are concerned with material planted and grown in fields, the term 'plots' is very common and often takes the place of 'individuals' or 'subjects'. However, we shall endeavour to maintain a general view, and will keep explicitly agricultural terminology to a minimum.

4.2 One-way arrangements

4.2.1 Introduction and model

We begin with the simplest situation, namely where the investigator wishes merely to compare the responses observed in several *a priori* groups. If there are just two such groups then we can use the simple two-sample t-test familiar from introductory statistics courses, but what can we do if there are more than two groups? For example, if

- responses are the times taken to obtain pain relief for subjects given one of six headache cures and we wish to determine if there is a difference between the cures in terms of average time to relief;
- responses are IQ scores from a standard test for students in five different schools and we wish to determine whether brighter students tend to go to a particular school.

One approach would be to take the groups two at a time, and to compare means by conducting all possible such two-sample tests. While this is clearly a feasible procedure, it is unsatisfactory and inefficient because:

(i) only the data from the pair of samples under consideration are used in each test, whereas often the complete set of data has been collected as a single experiment and *every* observation has a contribution to make towards calculation of such quantities as standard errors;
(ii) even with a moderate number of groups there will be many pairwise comparisons, and the more comparisons that are made the higher is the probability of obtaining false significant differences for some of the tests.

These arguments imply that we should use *all* the data and conduct just a *single* analysis of differences between the groups, so the way forward is to formulate an appropriate model for the whole situation.

Let us suppose that there are k *a priori* populations from which random samples of sizes $n_1, n_2, \ldots, n_k$ respectively have been taken, and that y_{ij} denotes the response for the jth individual in the sample from the ith population ($i = 1, \ldots, k; j = 1, \ldots, n_i$). Also, write $n = n_1 + \cdots + n_k$ for the total number of observations. Thus, following previous chapters, a simple model for the data is given by

$$y_{ij} = \mu_i + \epsilon_{ij}, \tag{4.1}$$

where μ_i is the mean of the ith population while ϵ_{ij} is a random variable indicating the jth individual's departure from the population mean ($i = 1, \ldots, k; j = 1, \ldots, n_i$).

All the foregoing examples suggest that when we say we are interested in finding differences between groups we really mean that we are interested in comparing the *means* of the underlying populations, so it is useful to impose some further structure on the above model. Let us suppose that μ is the overall (population) mean for the n individuals, so that $\mu = \frac{1}{n} \sum_i n_i \mu_i$. Then we may write

$$\mu_i = \mu + \alpha_i, \quad \text{where} \sum_i n_i \alpha_i = 0$$

and α_i is called the *effect* of population i. Frequently the populations correspond to different 'treatments', and it is then meaningful to talk of a *treatment effect*.

The model thus contains the systematic component $\mu + \alpha_i$ and the random component ϵ_{ij}. The final ingredient that is necessary is a specification of the distribution from which the ϵ_{ij} are drawn, and in common with previous models we will assume the mean of the ϵ_{ij} to be zero and the variance to be a constant σ^2 regardless of group of origin. Moreover, we will consider only the case of quantitative response variables in this chapter (for qualitative response variables see Chapter 7), so we will assume normality of the ϵ_{ij} for purposes of hypothesis testing and confidence interval construction. Thus, finally, the model is

$$y_{ij} = \mu + \alpha_i + \epsilon_{ij}, \tag{4.2}$$

where $\sum_i n_i \alpha_i = 0$ and the ϵ_{ij} are iid $N(0, \sigma^2)$ for $i = 1, \ldots, k$ and $j = 1, \ldots, n_i$.

4.2.2 Testing hypotheses

To test for differences among the k groups, we must look for differences among the k population means. Accordingly, we consider testing the null hypothesis

$$H_0 : \mu_1 = \mu_2 = \cdots = \mu_k = \mu$$

against the alternative that at least one of the μ_i differs from the rest, and this is exactly equivalent to testing the null hypothesis

$$H_0 : \alpha_1 = \alpha_2 = \cdots = \alpha_k = 0$$

against the alternative that at least one of the α_i is non-zero.

The general procedure for conducting this test is the analysis of variance. We have already met this procedure in Chapter 3 in the context of multiple regression, but it was originally developed within the framework of one-way experimental arrangements of the type being considered here. It is basically a procedure for partitioning a sum of squares into components associated with recognized sources of variation, so it lends itself to generalization in many directions and it is currently used in a very wide range of statistical applications (including, of course, regression).

We will require to sum and average quantities repeatedly throughout this chapter, so some shorthand notation will be convenient to avoid both overuse of the summation sign and possible confusion with overbars. To denote the mean of values over a particular classification we will replace the corresponding suffix in the lower-case letter by a dot, while the corresponding total will be similarly written but in an upper-case letter. Thus, in the present case, the mean of the ith group is $y_{i.} = \frac{1}{n_i} \sum_{j=1}^{n_i} y_{ij}$, the mean over all the groups (the 'grand mean') is $y_{..} = \frac{1}{n} \sum_{i=1}^{k} \sum_{j=1}^{n_i} y_{ij}$, the total in the ith group is $Y_{i.} = \sum_{j=1}^{n_i} y_{ij}$ and the grand total is $Y_{..} = \sum_{i=1}^{k} \sum_{j=1}^{n_i} y_{ij}$. (The same notation will be followed in subsequent sections, however many subscripts attach to each letter.)

To derive the test for differences among groups, we need to consider the (corrected) total sum of squares of the responses $\sum_{i=1}^{k} \sum_{j=1}^{n_i} (y_{ij} - y_{..})^2$. Adding and subtracting $y_{i.}$ within the square and suitably bracketing the terms, this sum of squares can be written as $\sum_{i=1}^{k} \sum_{j=1}^{n_i} [(y_{ij} - y_{i.}) + (y_{i.} - y_{..})]^2$. Then expanding the square in terms of the quantities in the round brackets, and remembering that $\sum_{j=1}^{n_i} (y_{ij} - y_{i.}) = 0$, it is easy to establish the identity

$$\sum_{i=1}^{k} \sum_{j=1}^{n_i} (y_{ij} - y_{..})^2 = \sum_{i=1}^{k} n_i (y_{i.} - y_{..})^2 + \sum_{i=1}^{k} \sum_{j=1}^{n_i} (y_{ij} - y_{i.})^2 \qquad (4.3)$$

The term on the left-hand side is the familiar total sum of squares of the responses, already denoted S_{yy} in Chapter 3. The first term on the right-hand side expresses the (weighted) squared differences among the group means, so is usually termed the *between-group* sum of squares; we will denote it SS_T (in recognition of the fact that the groups are often 'treatment groups'). The second term on the right-hand side can be recognized as the sum of squares pooled within the groups, so is called the *within-group* sum of squares; we will denote it SS_E (in recognition of the fact that it is a measure of the experimental error).

Thus the fundamental identity in one-way analysis of variance is

$$S_{yy} = SS_T + SS_E.$$

For computation purposes it is most convenient to re-express the first two sums of squares as

$$S_{yy} = \sum_{i=1}^{k} \sum_{j=1}^{n_i} y_{ij}^2 - ny_{..}^2$$

and

$$SS_T = \sum_{i=1}^{k} (Y_{i.}^2 / n_i) - Y_{..}^2 / n,$$

and to obtain SS_E by subtraction as $S_{yy} - SS_T$. (Note that $ny_{..}^2 = Y_{..}^2 / n$. This quantity is often referred to as the *correction factor*, as it 'corrects' the respective sum of squares for centering about the mean.)

Given model (4.2), the following distributional results can be readily derived:

(i) SS_E / σ^2 has a chi-squared distribution on $n - k$ degrees of freedom *always*;

(ii) if H_0 is true, SS_T / σ^2 has a chi-squared distribution on $k - 1$ degrees of freedom and is independent of SS_E;

(iii) if H_0 is not true, SS_T follows a non-central chi-squared distribution and $E(SS_T) = (k - 1)\sigma^2 + \sum_i n_i \alpha_i^2$.

Thus define the *mean squares* $MS_E = SS_E / (n - k)$ and $MS_T = SS_T / (k - 1)$. All these computations can be set out most conveniently in the analysis of variance (ANOVA) table:

Source of Variation	Sum of Squares	Degrees of Freedom	Mean Square	F-Ratio
Between groups	SS_T	$k-1$	MS_T	$F = \frac{MS_T}{MS_E}$
Within groups	SS_E	$n-k$	MS_E	
Total	S_{yy}	$n-1$		

The distributional results given earlier show that

(i) $E(MS_E) = \sigma^2$;

(ii) if H_0 is true then the ratio MS_T/MS_E has the F-distribution with $k-1$ and $n-k$ degrees of freedom, but if H_0 is not true then MS_T/MS_E has a non-central F-distribution with bigger mean.

The ratio $F = MS_T/MS_E$ can thus be used in conjunction with F-tables to conduct a formal test of hypothesis H_0, in the same manner as the ANOVA F-tests in Chapter 3 (i.e. we reject H_0 if the calculated value of F exceeds the critical value of the $F_{k-1,n-k}$ distribution at a given significance level).

Example 4.1. A manufacturer of tyres wishes to investigate their rate of wear, and whether this rate differs substantially among the four positions on the car that the tyre can occupy (1 = front offside, 2 = front nearside, 3 = rear offside and 4 = rear nearside). The manufacturer sets up an experiment in which the tyres are fitted to a car, the car is driven at a fixed speed for a fixed distance, and the reduction in depth of tread caused by the test is measured (in hundredths of a mm) for each tyre. The process is then repeated nine times with a new set of tyres each time. The results obtained are shown in Table 4.1.

From these results we obtain the following quantities.

(i) Group totals: $Y_{1.} = 194.593$, $Y_{2.} = 157.662$, $Y_{3.} = 293.204$, $Y_{4.} = 249.431$; grand total: $Y_{..} = 894.89$; grand mean: $y_{..} = 24.858$.

(ii) Total sum of squares: $S_{yy} = 20.935^2 + 19.013^2 + \cdots + 28.176^2 + 28.701^2 - 36 \times 24.858^2 = 2112.133$.

(iii) Between-group sum of squares: $SS_T = 194.593^2/9 + \cdots + 293.204^2/9 - 894.89^2/36 = 1189.014$.

(iv) Within-group sum of squares: $SS_E = 2112.133 - 1189.014 = 923.119$.

Hence we can draw up the ANOVA table:

Source	S.S.	d.f.	M.S.	F-Ratio
Between groups	1189.014	3	396.338	13.7
Within groups	923.119	32	28.847	
Total	2112.133	35		

Table 4.1 *Amount of wear of tyres in four car positions for fixed distance travelled.*

Position			
	20.935	17.123	29.590
1	19.013	15.919	28.092
	20.332	15.285	28.304
	18.279	14.815	19.973
2	21.200	11.280	20.096
	19.389	12.153	20.477
	28.535	37.227	30.529
3	27.998	38.853	29.177
	30.073	40.017	30.795
	20.182	34.340	29.023
4	18.792	34.707	28.176
	19.203	36.307	28.701

Critical values from the F-distribution on 3 and 32 degrees of freedom are 2.90 (for 5% significance), 4.47 (for 1% significance) and 7.00 (for 0.1% significance), so the calculated ratio of 13.7 clearly exceeds all of these critical values indicating evident difference among the four tyre positions.

4.2.3 Follow-up analysis

Preparing the ANOVA table and conducting the F-test for differences between groups is just the first stage of a full analysis of the data. Usually, one or more of the following aspects also needs to be considered.

Single degree-of-freedom contrasts

An F-test that has more than one degree of freedom for the numerator mean square is an average of as many independent comparisons as there are degrees of freedom. While the null hypothesis that all population means are equal is specified uniquely, the alternative that at least one differs from the rest can be satisfied in many different ways. Often, an experiment is planned to provide answers to a number of separate questions (e.g. does the 'control' treatment exert a different effect from the average of all the other treatments? Do two specified treatments have different effects? Do two sets of treatments have different effects on average?) In such circumstances the overall F-test will be rather unhelpful, as it will be directed at a mixture of all these questions. We need to disentangle the answers to the individual questions from the overall result, and this can be done by partitioning the sum of squares between groups into single-degree-of-freedom components. However, it is important to recognize that for validity of analysis these combinations need to be specified *before* the data have been collected; we return to this point below.

Any linear combination of group totals $z_w = l_{w1}Y_1 + l_{w2}Y_2 + \cdots + l_{wk}Y_k$. is called a *linear comparison*, or *linear contrast*, among the group totals if the coefficients satisfy the condition $\sum_{i=1}^{k} n_i l_{wi} = 0$, and two contrasts z_1 and z_2 are said to be *orthogonal* if $\sum_{i=1}^{k} n_i l_{1i} l_{2i} = 0$. Thus, for example, if there are three groups with equal sample sizes $n_1 = n_2 = n_3 = m$, say, such that group 1 is a 'control' while groups 2 and 3 correspond to two new experimental regimes A and B, then $z_1 = Y_1 - \frac{1}{2}(Y_2 + Y_3)$ is a contrast between the control group and the two experimental regimes, $z_2 = Y_2 - Y_3$ is a contrast between the two experimental regimes, and these two contrasts are orthogonal.

We can partition the between-group sum of squares SS_T using the following results.

1. If $z_w = \sum_i l_{wi} Y_i$ is *any* contrast among the Y_i and $D_w = n_1 l_{w1}^2 + n_2 l_{w2}^2 + \cdots + n_k l_{wk}^2$, then the quantity z_w^2 / D_w is a component of SS_T representing one degree of freedom.

2. If z_1 and z_2 are *orthogonal*, then z_2^2 / D_2 is similarly a one-degree-of-freedom component of $SS_T - z_1^2 / D_1$.

3. If SS_T has $k-1$ degrees of freedom and $z_1, z_2, \ldots, z_{k-1}$ are (any) *mutually orthogonal* contrasts, then

$$SS_T = \frac{z_1^2}{D_1} + \frac{z_2^2}{D_2} + \cdots + \frac{z_{k-1}^2}{D_{k-1}}.$$

Starting with a specified z_1, we can *always* find appropriate $z_2, z_3, \ldots, z_{k-1}$ to construct a completely orthogonal set (in fact, the greater the value of k, the greater is the number of distinct possibilities). At any stage there will in general be several possible choices of contrast orthogonal to all preceding ones, and the analyst must choose the most relevant set for interpretation. Since each contrast z_w has one degree of freedom the quantity z_w^2 / D_w is also a mean square, and the corresponding effect can thus be tested for significance by referring the ratio $(z_w^2 / D_w) \div MS_E$ to the $F_{1,n-k}$ distribution. (Note, also, that if *non-orthogonal* contrasts are selected by mistake, then the sum of their z_w^2 / D_w values will not equal SS_T).

Example 4.1 (Cont.). Suppose the manufacturer is interested in the average effects on wear of front compared with rear tyres, and of nearside compared with offside positions. We can partition the between-group sum of squares in such a way that these effects are obtained as orthogonal contrasts with a single degree of freedom each. Consider

$$z_1 = Y_1 + Y_2 - Y_3 - Y_4 = -190.38$$

and

$$z_2 = Y_{1.} - Y_{2.} + Y_{3.} - Y_{4.} = 80.704$$

Then z_1 contrasts front and rear tyres, z_2 contrasts offside and nearside, and z_1 and z_2 are orthogonal. We also have $D_1 = 9 \times 1 + 9 \times 1 + 9 \times 1 + 9 \times 1 = 36$, and similarly $D_2 = 36$. Hence the component of SS_T due to 'front–rear' effect is $SS_{FR} = (190.38)^2/36 = 1006.79$, while the component due to 'offside–nearside' effect is $SS_{ON} = 80.704^2/36 = 180.92$. These two components have 1 d.f. each, so are also the respective mean squares. The significance of each mean square is tested by dividing it by MS_E and referring the result to the F-distribution on 1 and 32 degrees of freedom. We obtain the ratios $1006.79 \div 28.847 = 34.9$ and $180.92 \div 28.847 = 6.27$ respectively. The 5% significance level critical value of the $F_{1,32}$ distribution is 4.15, so both 'front–rear' and 'offside–nearside' contrasts show a significant effect at this level. Moreover, there is still one degree of freedom left after these effects have been removed from SS_T. This contrast must have a sum of squares equal to $1189.014 - 1006.79 - 180.92 = 1.304$, so evidently is not significant. The explanation for the observed difference between the four groups thus resides exclusively in the two contrasts given above.

Point estimation

To estimate parameters in the model (4.2), we can either use maximum likelihood or we can relax the assumption of normality of departures and use least squares. Both approaches lead to the minimization of

$$V = \sum_{i=1}^{k} \sum_{j=1}^{n_i} (y_{ij} - \mu - \alpha_i)^2$$

for estimation of μ and α_i ($i = 1, \ldots, k$). Differentiating V with respect to each of these quantities in turn, setting the resulting expressions to zero and solving the ensuing $k + 1$ simultaneous equations (remembering in the process that $\sum_i n_i \alpha_i = 0$), we obtain the estimates

$$\hat{\mu} = y_{..} \quad \text{and} \quad \hat{\alpha}_i = y_{i.} - y_{..} \quad (i = 1, \ldots, k). \tag{4.4}$$

The Gauss–Markov theorem then tells us that the best linear unbiased estimator of $\sum_i l_i \alpha_i$ is $\sum_i l_i \hat{\alpha}_i$ for any set of coefficients l_i, and the best estimate of σ^2 is MS_E from the ANOVA table. This latter is thus often written s_e^2 and called the *residual mean square*.

Confidence intervals

Interest most often focusses either on individual group effects (i.e. the α_i) or linear combinations/contrasts of group effects, and it is generally useful to obtain a confidence interval for the quantity of interest. In this case we must include the assumption of normality of departures in our model, and we invoke the result on linear combinations of independent normal variables (Section 2.4) to deduce that

$$\sum_{i=1}^{k} a_i y_{i.} \sim N\left(\sum_{i=1}^{k} a_i \alpha_i, \sigma^2 \sum_{i=1}^{k} \frac{a_i^2}{n_i}\right) \tag{4.5}$$

for any specified set of constants a_i. Hence, following standard confidence interval derivation (Section 2.5.2), we deduce that a $100(1 - \alpha)\%$ confidence interval for $\sum_{i=1}^{k} a_i \alpha_i$ is given by

$$\sum_{i=1}^{k} a_i y_{i.} \pm t_{n-k,\alpha} s_e \sqrt{\sum_{i=1}^{k} \frac{a_i^2}{n_i}}, \tag{4.6}$$

where $t_{n-k,\alpha}$ is the critical value from tables such that $\Pr(-t_{n-k,\alpha} \le T \le +t_{n-k,\alpha}) = 1 - \alpha$ when T has a t-distribution on $n - k$ degrees of freedom. (Note that the number of degrees of freedom in this t multiplier is the number in the 'within-groups' line of the ANOVA, as this is the number on which the estimate s_e^2 of σ^2 is based). This expression is a very general one, and the confidence interval for any individual α_i, or for any difference of two α_i, can be obtained from it by appropriate choice of a_i.

Example 4.1 (Cont.). Returning again to the tyre-wear example above, the estimated standard error of a single treatment mean is $\sqrt{28.847/9} = 1.79$, that of a difference between two treatment means is $\sqrt{2 \times 28.847/9} = 2.53$, and so on. These standard errors are then used in confidence interval calculations. For example, a 95% confidence interval for the difference in mean wear between front offside and front nearside is $(y_{1.} - y_{2.}) \pm t_{32,0.95} \times 2.53$. We have $y_{1.} = 21.621$, $y_{2.} = 17.518$ and $t_{32,0.95} = 2.038$, so the interval is 4.10 ± 5.16, i.e. $(-1.06, 9.26)$.

Least significant difference

One particular planned contrast that can be tested for significance by means of the single-degree-of-freedom theory above is a simple difference between two means. In this case, following through the algebra establishes that the F-ratio for the contrast is equal to the square of the two-sample t-statistic for testing the difference in means using the pooled error variance s_e^2. The ANOVA test is thus exactly equivalent to a t-test on the means. Often, several such differences may be of interest to the experimenter and then we would conduct separate t-tests on each difference.

If the number of individuals is the same in each group, i.e. $n_i = r$ for $i = 1, \ldots, k$ so that $r = n/k$, then a convenient quantity to calculate for testing hypotheses about such differences is the *least significant difference (lsd)*. The lsd for a test at significance level $100\alpha\%$ is given by $s_e t_{n-k,\alpha} \sqrt{2/r}$. For a difference between two treatment means to be significant at this level, the observed difference must exceed the lsd value. This is just an alternative way of stating the two-sample t-test with pooled error variance s_e^2.

This value need only be calculated once, so is a timesaver compared with making individual t-tests. However, it can easily be misused so care needs to be exercised when employing it. It is only valid for *planned* comparisons between treatments, and should never be used to make comparisons *suggested by the data* or to identify potentially different treatments. In particular, comparing the largest and smallest means can be a great temptation for the naive experimenter but will always lead to gross distortions. For example, when testing at the 5% level of significance and comparing the largest and smallest means in this way, the lsd is exceeded on 13% of occasions when 3 independent groups have been taken from a single population, on 40% of occasions when 6 independent groups have been taken from a single population, and on 90% of occasions when 20 independent groups have been taken from a single population! These results show the dangers of making such 'post-hoc' comparisons, as in none of these cases is there a 'real' difference between any of the groups.

4.3 Cross-classifications

4.3.1 Two-way arrangements with additive effects: randomized blocks

The methods of the previous section are appropriate when no sources of variation other than group effects are anticipated. In particular, it is assumed that the different individuals in each group are subject to the same systematic group influence but otherwise any differences between them are purely random effects. However, in many situations it is known beforehand that certain individuals are likely to behave more similarly than other individuals. For example:

(i)   adjacent plots in agricultural field experiments will usually be more alike in response than ones far apart;

(ii) animals from the same litter will generally show more similarities than ones from different litters;

(iii) observations made on the same day, or using a particular piece of equipment, may resemble each other more than those made on different days or using different pieces of equipment.

There are two implications for such situations where the behaviour of individual units may be anticipated in part and the units classified accordingly. One concerns the *design* of the study: it is important to arrange the experimental material in such a way as to make optimal use of this partial information, and to prevent the anticipated systematic effects of the units from interfering with the objectives of the study. This aspect is a major consideration in the *planning* of statistical investigations, and is outside the scope of this book. We will concentrate instead on the second implication, concerning the *analysis* of the results: we need to incorporate these anticipated effects within the systematic part of the model so that they are excluded from the estimated experimental error and the analysis is thereby made more precise.

Consider again the data in Example 4.1, and suppose that we are told by the investigator that the nine 'repeats' of the experiment had actually been conducted in three batches of three, and that each batch of three repeats used a different car for the trials. The data in the three columns of Table 4.1 are the results using cars A, B and C respectively.

Given this information we might wish to reconsider our analysis of the trials, because we would (in general) expect a systematic difference between the three cars as regards the amount of tyre wear that they induce (due to different car weights, roadholding, balance, and so on). Each value in the table can now be categorized in two ways: according to which tyre position it corresponds to (i.e. the *rows*) and according to which car it came from (i.e. the *columns*), and *both* of these categorizations need to be taken into account in the analysis. An alternative way of viewing the data is to say that the responses have been collected into three *blocks* (i.e. the cars), and within these blocks each of the former groups (i.e. the tyre positions) is represented by independent observations. Consequently this arrangement is often known as a *randomized blocks* design. We now outline the analysis of such a design, but stress that there is one vital condition that has to be satisfied in this analysis: the number of independent repeats of each group *must be the same in each block*. If this condition is not satisfied then the general method of Section 4.5 needs to be used.

Models and assumptions

Let us suppose, as before, that there are k groups forming the primary focus of the investigation, but that the n experimental units are now divided among b blocks with r observations from each group within each block (so that $n = kbr$). Then denote by y_{iju} the observation on the uth individual belonging to group i in block j ($i = 1, \ldots, k$; $j = 1, \ldots, b$; $u = 1, \ldots, r$). Since we anticipate systematic differences between the blocks, let us assume that the jth block has some *effect* β_j for $j = 1, \ldots, b$, to go with the group effects α_i ($i = 1, \ldots, k$) already postulated in the previous analysis. The simplest way of incorporating the extra information into the previous model (4.2) is to assume that each block will have a constant systematic effect on an experimental unit, regardless of which group that unit belongs to. This implies that the block and group effects are *additive*, so that model (4.2) is modified to

$$y_{iju} = \mu + \alpha_i + \beta_j + \epsilon_{iju}, \tag{4.7}$$

where the ϵ_{iju} are iid $N(0, \sigma^2)$ for $i = 1, \ldots, k$; $j = 1, \ldots, b$ and $u = 1, \ldots, r$. Since we now have an equal number of individuals in each group, the constraint on the group effects simplifies to $\sum_i \alpha_i = 0$, and because of the presence of μ we need a corresponding constraint $\sum_j \beta_j = 0$ on the block effects.

Testing hypotheses

Once again the prime interest is to test for differences among the k groups, so we consider the null hypothesis

$$H_{0a} : \alpha_1 = \alpha_2 = \cdots = \alpha_k = 0$$

against the alternative that at least one of the α_i is non-zero. Although the block effects β_j are rarely of interest in their own right and we generally just wish to eliminate them from the experimental error, it is nevertheless possible to formulate the analogous hypothesis

$$H_{0b} : \beta_1 = \beta_2 = \cdots = \beta_b = 0$$

with the alternative that at least one of the β_j is non-zero.

These hypotheses are tested once again by analysis of variance, and we start by partitioning the total sum of squares in a meaningful way. This time we have means $y_{i..}$ (of the rb observations in the ith group), $y_{.j.}$ (of the rk observations in the jth block), $y_{ij.}$ (of the r observations from group i in block j), and $y_{...}$ (of all rbk observations), with corresponding totals $Y_{i..}$, $Y_{.j.}$, $Y_{ij.}$ and $Y_{...}$. Then if each of these means is added to and subtracted from $y_{iju} - y_{...}$ and terms are bracketed appropriately we obtain

$$(y_{iju} - y_{...}) = (y_{i..} - y_{...}) + (y_{.j.} - y_{...}) + (y_{iju} - y_{i..} - y_{.j.} + y_{...}). \quad (4.8)$$

Squaring both sides of this equation and summing over suffixes i, j and u, all cross-product terms vanish (because they all include at least one sum of values about their mean) and we obtain the identity

$$S_{yy} = SS_T + SS_B + SS_E, \quad (4.9)$$

where

$S_{yy} = \sum_i \sum_j \sum_u (y_{iju} - y_{...})^2$, the usual total (corrected) sum of squares,

$SS_T = br \sum_i (y_{i..} - y_{...})^2$, the sum of squares between groups,

$SS_B = kr \sum_j (y_{.j.} - y_{...})^2$, the sum of squares between blocks, and

$SS_E = \sum_i \sum_j \sum_u (y_{iju} - y_{i..} - y_{.j.} + y_{...})^2$, the residual (or error) sum of

squares.

For computation purposes it is simpler to obtain these sums of squares as:

$$S_{yy} = \sum_{i=1}^{k} \sum_{j=1}^{b} \sum_{u=1}^{r} y_{iju}^2 - ny_{...}^2,$$

$$SS_T = \sum_{i=1}^{k} (Y_{i..}^2/br) - Y_{...}^2/n,$$

$$SS_B = \sum_{j=1}^{b} (Y_{.j.}^2/kr) - Y_{...}^2/n,$$

and SS_E by subtraction $(S_{yy} - SS_T - SS_B)$.

Given model (4.7), the following distributional results can be readily derived:

(i) SS_E/σ^2 has a chi-squared distribution on $n-k-b+1$ degrees of freedom *always*;

(ii) *if H_{0a} is true*, SS_T/σ^2 has a chi-squared distribution on $k-1$ degrees of freedom and is independent of SS_B and SS_E;

(iii) *if H_{0a} is not true*, SS_T follows a non-central chi-squared distribution and $E(SS_T) = (k-1)\sigma^2 + b\sum_i \alpha_i^2$.

(iv) *if H_{0b} is true*, SS_B/σ^2 has a chi-squared distribution on $b-1$ degrees of freedom and is independent of SS_T and SS_E;

(v) *if H_{0b} is not true*, SS_B follows a non-central chi-squared distribution and $E(SS_B) = (b-1)\sigma^2 + k\sum_j \beta_j^2$.

Thus define the mean squares $MS_E = SS_E/(n-k-b+1)$, $MS_T = SS_T/(k-1)$ and $MS_B = SS_B/(b-1)$, and set out the calculations in the ANOVA table:

Source of Variation	Sum of Squares	Degrees of Freedom	Mean Square	F-Ratio
Between blocks	SS_B	$b-1$	MS_B	$F_B = \frac{MS_B}{MS_E}$
Between groups	SS_T	$k-1$	MS_T	$F_T = \frac{MS_T}{MS_E}$
Error	SS_E	$n-k-b+1$	MS_E	
Total	S_{yy}	$n-1$		

The distributional results given earlier show that

(i) $E(MS_E) = \sigma^2$;

(ii) if H_{0a} is true then the ratio F_T has the F-distribution with $k - 1$ and $n - k - b + 1$ degrees of freedom, but if H_{0a} is not true then F_T has a non-central F-distribution with bigger mean.

(iii) if H_{0b} is true then the ratio F_B has the F-distribution with $b - 1$ and $n - k - b + 1$ degrees of freedom, but if H_{0b} is not true then F_B has a non-central F-distribution with bigger mean.

The ratios F_T and F_B can thus be used in conjunction with F-tables to conduct formal tests of the hypotheses H_{0a} and H_{0b} in the same manner as all previous ANOVA F-tests, the null hypothesis in each case being rejected if the calculated ratio exceeds the critical value from the tables.

Single degree-of-freedom contrasts

The sum of squares between groups can once again be partitioned into contrasts among the group totals representing a single degree of freedom each. The results given in Section 4.2.3 for one-way arrangements apply exactly as before, the only points to remember being that the group totals are now denoted by $Y_{i..}$ for $i = 1, \ldots, k$ and that each group contains the same number of observations (so that all n_i are equal to rb in each D_w).

Analagous contrasts could, if desired, be defined for the block totals $Y_{.j.}$ and the sum of squares between blocks could be partitioned similarly into single-degree-of-freedom contrasts. However, in most applications the purpose of incorporating blocks in the analysis is to make comparisons between the groups more precise, and investigation of differences between blocks is rarely of interest in its own right. Consequently partitioning SS_B in this way is rarely undertaken.

Point estimation

Application of either maximum likelihood or least squares to estimate the parameters μ, α_i and β_j in model (4.7) leads to minimization of

$$V = \sum_{i=1}^{k} \sum_{j=1}^{b} \sum_{u=1}^{r} (y_{iju} - \mu - \alpha_i - \beta_j)^2.$$

Straightforward calculus yields the estimates

$$\hat{\mu} = y_{...}, \quad \hat{\alpha}_i = y_{i..} - y_{...} \quad \text{and} \quad \hat{\beta}_j = y_{.j.} - y_{...} \tag{4.10}$$

for $i = 1, \ldots, k$ and $j = 1, \ldots, b$. As before, the estimate s_e^2 of σ^2 in (4.7) is given by the error mean square MS_E from the ANOVA table.

Confidence intervals

Analogously to the results in Section 4.2.3 for the one-way arrangement, we have the results

$$\sum_{i=1}^{k} a_i y_{i..} \sim N\left(\sum_{i=1}^{k} a_i \alpha_i, \frac{\sigma^2}{br} \sum_{i=1}^{k} a_i^2\right) \tag{4.11}$$

and

$$\sum_{j=1}^{b} a_j y_{.j.} \sim N\left(\sum_{j=1}^{b} a_j \beta_j, \frac{\sigma^2}{kr} \sum_{j=1}^{b} a_j^2\right) \tag{4.12}$$

for constants a_i, a_j. Thus a $100(1-\alpha)\%$ confidence interval for $\sum_{i=1}^{k} a_i \alpha_i$ is given by

$$\sum_{i=1}^{k} a_i y_{i..} \pm t_{n-k-b+1,\alpha} s_e \sqrt{\frac{1}{br} \sum_{i=1}^{k} a_i^2}, \tag{4.13}$$

where $t_{n-k-b+1,\alpha}$ is the critical value from tables such that $\Pr(-t_{n-k-b+1,\alpha} \le T \le +t_{n-k-b+1,\alpha}) = 1 - \alpha$ when T has a t-distribution on $n - k - b + 1$ degrees of freedom. (Note again that the number of degrees of freedom in this t multiplier is the number in the 'error' line of the ANOVA, as this is the number on which the estimate s_e^2 of σ^2 is based).

A corresponding $100(1-\alpha)\%$ confidence interval for $\sum_{j=1}^{b} a_j \beta_j$ is given by

$$\sum_{j=1}^{b} a_j y_{.j.} \pm t_{n-k-b+1,\alpha} s_e \sqrt{\frac{1}{kr} \sum_{j=1}^{b} a_j^2}, \tag{4.14}$$

but as already discussed above we are rarely interested in such comparisons among blocks *per se*.

Example 4.2. Let us suppose that indeed three different cars were used for the tyre-wear trials, and that the results in the first column of Table 4.1 relate to car A, those in the middle column to car B, and those in the end column to car C. We wish to eliminate any systematic 'car' effect from comparisons among the four tyre positions, so must use the randomized block analysis described above. We have $k = 4$, $b = 3$ and $r = 3$ repeats in each car/position combination.

The group totals (194.539, 157.662, 293.204, 249.431), grand total (894.89) and grand mean (24.858) are obviously unchanged, as also are the total sum of squares (2112.133) and the between-group sum of squares (1189.014). The new calculations needed for the present analysis are:

(i) Block (i.e. car) totals: $Y_{.1.} = 263.931$, $Y_{.2.} = 308.026$ and $Y_{.3.} = 322.970$;
(ii) Between-blocks sum of squares: $SS_B = 263.931^2/12 + 308.026^2/12 + 322.970^2/12 - 894.89^2/36 = 156.884$;
(iii) Error sum of squares: $SS_E = 2112.133 - 1189.014 - 156.884 = 766.235$.

Thus we have the new ANOVA table:

Source	S.S.	d.f.	M.S.	F-Ratio
Between blocks	156.884	2	78.442	3.07
Between groups	1189.014	3	396.338	15.5
Error	766.235	30	25.541	
Total	2112.133	35		

We note that removing the block effects has reduced the error mean square from 28.847 in Example 4.1 to 25.541, and this has sharpened up the F-test for differences between positions by increasing the calculated ratio from 13.7 to 15.5. (Note, however, that the 2 degrees of freedom attributable to blocks have been removed from those available for error, so the latter have fallen from the previous 32 to 30). The critical values of the $F_{2,30}$ distribution are 2.49 (for 10% significance) and 3.32 (for 5% significance). Thus the calculated ratio $F_B = 3.07$ indicates that there is *some* effect due to the different cars used, albeit not a very strong one. Nevertheless, there are sufficient systematic differences in tyre wear between the different cars to justify the randomized blocks analysis. Since twelve observations make up each block, the standard error of a block mean is $\sqrt{\frac{25.541}{12}}$ and that of a treatment mean is $\sqrt{\frac{25.541}{9}}$. Standard errors of contrasts or differences are calculated in the usual way and lead to confidence intervals. The sum of squares between groups breaks down into single-degree-of-freedom contrasts exactly as in Example 4.1, but the new error mean square enhances the F-ratios for these contrasts slightly.

4.3.2 Interaction

Model (4.7) allows for systematic block effects when investigating differences between groups, but it does so in a somewhat restricted way. The fundamental assumption in the model is that each block has *the same* effect on every group, i.e. that the block and group effects are additive. Unfortunately, such a situation is often an oversimplification.

For example, consider grouping subjects in a psychological experiment by age. Suppose that the experiment involves reacting to various stimuli (e.g. different coloured lights), the response variable is the speed of the subject's reaction, and we treat the different age groups as blocks for the purpose of the analysis. It is certainly true that different age groups may react differently to the stimuli (generally, the older the subject the slower will be his or her reaction), so blocking subjects by age is a sensible measure that will improve precision of analysis. However, it is by no means certain that the 'age differences' will be constant over all stimuli. Some colours will be more visible than others, and we would therefore expect the slowing reaction with age to be less pronounced for these colours than for the others. In similar vein, we might expect differential effects of litters in responses to diets in animal feed trials, differential effects of panellists in food-tasting experiments, and so on. We therefore need to modify the previous analysis if we are to deal satisfactorily with such situations.

Returning to the notation of the previous section, we again let y_{iju} denote the observation on the uth individual belonging to group i in block j ($i = 1, \ldots, k$; $j = 1, \ldots, b$; $u = 1, \ldots, r$), but now simply express the model as

$$y_{iju} = v_{ij} + \epsilon_{iju}, \tag{4.15}$$

where the ϵ_{iju} are iid $N(0, \sigma^2)$. Thus v_{ij} represents the true mean of individuals in group i and block j.

If groups and blocks *are* additive then, from (4.7),

$$v_{ij} = \mu + \alpha_i + \beta_j.$$

Plotting the v_{ij} will yield *parallel lines*, as shown in the example of Figure 4.1(a) where $k = 3$ and $b = 3$. However, if groups and blocks are *not* additive then the plots will depart from parallelism. Two possible types of departure are divergence of lines, as in Figure 4.1(b) where the differences between groups are always in the same direction but increase progressively from block 1 to block 3, or crossing-over of lines, as in Figure 4.1(c) where the pattern of differences between groups changes between blocks.

In these latter cases we need an extra term in the model, so let us write

$$v_{ij} = \mu + \alpha_i + \beta_j + \gamma_{ij}$$

where γ_{ij} measures the extent of departure from additivity. We say that the groups *interact* with the blocks, or that there is a *group by block interaction*.

When we consider data from such a cross-classification, the means $y_{ij.}$ for each group/block combination estimate the v_{ij} ($i = 1, \ldots, k$; $j = 1, \ldots, b$), so plotting the $y_{ij.}$ in the manner of Figure 4.1 may show up the presence of interaction. However, sampling variability will now play a part, and considerable visual deviation from parallelism of lines can still be consistent with additivity of effects. We cannot therefore rely on such simple graphical devices, and need a formal test for the presence of interaction. The following gives the justification for such a test, derived from the ANOVA of the previous section. Note, however, that it can only be used if the number of observations (r) in each group/block combination is *greater than* 1.

First, note that there are bk 'combinations', each containing r observations. If we treat these combinations as groups in a one-way arrangement, then the methods of Section 4.2 show that the sum of squares 'between combinations' is given by

$$SS_C = \sum_{i=1}^{k} \sum_{j=1}^{b} Y_{ij.}^2/r - Y_{...}^2/n$$

while the sum of squares 'within combinations' (i.e. error) is

$$SS_E = S_{yy} - SS_C.$$

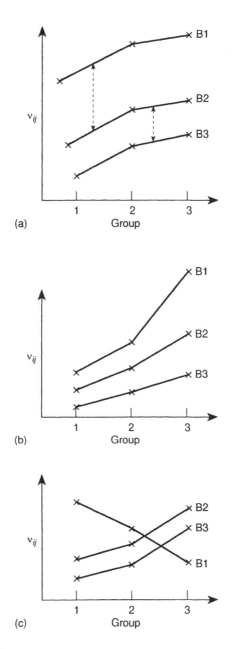

Fig. 4.1 *Possible patterns of means of responses between groups over blocks.*

Now if groups and blocks *are* additive, then the sum of squares between combinations must be recoverable from the randomized blocks ANOVA as $SS_T + SS_B$. Hence the difference

$$SS_I = SS_C - SS_T - SS_B$$

is the 'extra sum of squares' due to non-additivity of groups and blocks, and is known as the *interaction* sum of squares. Furthermore, since there are $bk - 1$ degrees of freedom for SS_C, $b - 1$ degrees of freedom for SS_B and $k - 1$ degrees of freedom for SS_T, there are $(bk-1)-(b-1)-(k-1) = (b-1)(k-1)$ degrees of freedom for SS_I so the corresponding mean square is $MS_I = SS_I/(b-1)(k-1)$. We can thus draw up the ANOVA table:

Source of Variation	Sum of Squares	Degrees of Freedom	Mean Square	F- Ratio
Between blocks	SS_B	$b - 1$	MS_B	$F_B = \frac{MS_B}{MS_E}$
Between groups	SS_T	$k - 1$	MS_T	$F_T = \frac{MS_T}{MS_E}$
Interaction	SS_I	$(b - 1)(k - 1)$	MS_I	$F_I = \frac{MS_I}{MS_E}$
Error	$SS_E (= S_{yy} - SS_C)$	$n - bk$	MS_E	
Total	S_{yy}	$n - 1$		

Distribution theory pertaining to the quantities in this table can be obtained in the same way as for previous tables, so results are not given here but may be assumed to form the basis for testing the effects in each line of the ANOVA. The tests are conducted in the usual manner, by referring the calculated ratios of mean squares to F-tables on the appropriate numbers of degrees of freedom. In particular, the test for interaction uses the ratio F_I. If this test is significant, it establishes non-additivity of effects. In this case the group means vary with blocks, so it does not make sense to average over blocks when reporting group effects and we need to report conclusions separately for the different blocks.

Example 4.3. Returning to the tyre example, the following are the totals of the three observations in each position/car combination:

Position	Car A	Car B	Car C
1	60.280	48.327	85.986
2	58.868	38.248	60.546
3	86.606	116.097	90.501
4	58.177	105.354	85.900

From these values we thus obtain

(i) $SS_C = (60.280^2 + 48.327^2 + \cdots + 105.354^2 + 85.900^2)/3 - 894.89^2/36 = 2084.549$,

(ii) $SS_I = 2084.549 - 1189.014 - 156.884 = 738.651$,

(iii) $SS_E = 2112.133 - 2084.549 = 27.584$,

so that we can draw up the new ANOVA table

Source	S.S.	d.f.	M.S.	F-Ratio
Between blocks	156.884	2	78.442	68.25
Between groups	1189.014	3	396.338	344.84
Interaction	738.651	6	123.109	107.11
Error	27.584	24	1.149	
Total	2112.133	35		

There has been a dramatic reduction in the error sum of squares from its value in Example 4.2, the decrease being explained by the fact that most of the previous sum of squares is attributable to the interaction sum of squares. The consequence is that all three calculated ratios are very large and the corresponding effects ('Positions', 'Cars' and 'Position by Car interaction') are very highly significant. The highly significant interaction means that the average responses vary considerably between the 12 position/car combinations, as can be seen clearly in the table of totals above. This table is thus the appropriate summary to quote for any conclusions, as it would be very misleading to average such diverse values over either positions or cars. We will explore the reasons behind these results in subsequent examples, but before we can do so we must discuss the idea of factorial treatment arrangements.

4.3.3 Factorial arrangements

So far we have considered how to investigate differences between groups, and we have allowed for the possibility that the individuals are blocked in some way. The analyses developed have been general ones, and no connections have been assumed between the groups. In practice, however, we often encounter structure among the groups. In particular, they are often combinations of two or more *factors*, each of which is represented at two or more *levels*. In the tyre wear example, for instance, the four groups represent all possible combinations formed from one of 'front' or 'rear' with one of 'nearside' or 'offside'. The first two categories may thus be treated as levels of one factor ('direction', say) while the last two as levels of a second factor ('side', say), and the objective of the analysis is to determine which of these factors (if either) has a significant effect on tyre wear. As another example, consider a medical researcher who wishes to study the effects of a certain diet, a certain exercise regime, and a certain drug in treating a particular condition. There are thus three factors (diet, exercise and medicine), and each can be either present or absent so has two levels. Denote presence of diet by D and absence by ND; presence of exercise by E and absence by NE; and presence of medicine by M and absence by NM. A factorial experiment would then consist of creating 8 groups of individuals, subjecting each to one of the 8 possible combinations of levels of the 3 factors (i.e. D/E/M, D/E/NM, D/NE/M, ND/E/M, D/NE/NM, ND/E/NM, ND/NE/M and ND/NE/NM) and analysing the responses to see which of the factors had a significant effect in relieving the condition.

Some basic concepts

We will denote factors by capital letters $(A, B, \ldots)$, levels of a factor by combinations of lower case letters and suffixes (e.g. a_1 is the first level of A, b_3 is the third level of B, etc), and either factor combinations or means by combinations of such symbols (e.g. $a_1 b_3$ refers to the group in which factor A is at level 1 and factor B is at level 3, and also denotes the mean response of this group). An experiment which has k factors at levels $r_1, r_2, \ldots, r_k$ respectively is called an $r_1 \times r_2 \times \cdots \times r_k$ experiment.

The basic ideas are most easily outlined with reference to a 2×2 factorial experiment. Denote the factors by A, B, the levels by a_1, a_2, b_1, b_2 and the four response means (over all replicates) by $a_1 b_1, a_1 b_2, a_2 b_1$ and $a_2 b_2$.

The two differences $a_2 b_1 - a_1 b_1$ and $a_2 b_2 - a_1 b_2$ are the *simple effects of A*: they give the difference in average response to the different levels of A at each level of B. The corresponding simple effects of B are $a_1 b_2 - a_1 b_1$ and $a_2 b_2 - a_2 b_1$.

The average of all simple effects of a factor is called the *main effect* of that factor. Thus the main effect of A is

$$M_A = \frac{1}{2}[(a_2 b_2 - a_1 b_2) + (a_2 b_1 - a_1 b_1)],$$

and the main effect of B is

$$M_B = \frac{1}{2}[(a_2 b_2 - a_2 b_1) + (a_1 b_2 - a_1 b_1)].$$

If both simple effects are the same for a factor, they will equal the main effect. In such a situation the effect of A is the same *whatever the level of B* and vice-versa, so the effects of A and B are *additive*. If simple effects differ for a factor then the effect of that factor differs across the levels of the other factor(s), so there is an *interaction* between the factors. The interaction is measured as the difference between the simple effects:

$$I_{AB} = \frac{1}{2}[(a_2b_2 - a_1b_2) - (a_2b_1 - a_1b_1)],$$

the value $\frac{1}{2}$ being used to maintain the same scale as the main effects. Interaction is thus a measure of departure of simple effects from additivity, so interpretation is exactly as discussed for randomized blocks and illustrated in Figure 4.1.

To illustrate these quantities, suppose the following table represents the means over all replicates in a 2×2 factorial experiment:

	a_1	a_2
b_1	30	32
b_2	36	44

Then it can be readily verified that $M_A = 5$, $M_B = 9$ and $I_{AB} = 3$.

When factors have more than two levels, main effects and interactions will involve more than one independent comparison. For example, if a factor has three levels then it will give rise to three simple effects and there are always two independent comparisons that can be made between three quantities. The main effect of the factor will therefore encompass any two such independent comparisons that can be found. Similarly, the interaction between this factor and any other one will involve (at least) three differences to be compared, and so will comprise at least three independent comparisons.

Analysis of a two-factor arrangement

Of course, the presence of main effects or interactions cannot be detected simply by inspecting means, as the observations are subject to chance fluctuations. We must determine whether or not the values obtained are greater than can be attributed to chance alone. To test for significance of effects we go back to the basic analysis of variance already developed, and we partition the sum of squares for groups into components representing main effects and interactions.

In the 2×2 factorial arrangement illustrated above there are 3 degrees of freedom for groups, which can be partitioned into 1 degree of freedom for each main effect and 1 for interaction. In a general 2-factor $p \times q$ arrangement, however, there are $pq - 1$ degrees of freedom for groups which yield $p - 1$ degrees of freedom for the main effect of the first factor, $q - 1$ for the main effect of the second factor, and $(p - 1)(q - 1)$ for interaction.

The details are as follows. We suppose that factor P has p levels, factor Q has q levels, there are r observations at each combination of factor levels, y_{ijk} is the kth observation at level i of P and level j of Q, and totals are denoted as usual by $Y_{i..}$ (for level i of P over all other classifications), $Y_{.j.}$ (for level j of Q over all other classifications) and $Y_{ij.}$ (for the r values at level i of P and level j of Q). Then:

(i) Compute the analysis of variance without regard to the factorial structure but taking due account of blocks if present. (If blocks are present, and there are b of them, then let $r = bs$). This yields the sums of squares for blocks (SS_B), groups (SS_T), error (SS_E) and total (S_{yy}), as well as the correction factor $C = Y^2_{...}/pqr$.

(ii) From the various totals, compute the sums of squares for main effects as

$$SS_P = \frac{1}{qr} \sum_{i=1}^{p} Y^2_{i..} - C$$

for the main effect of P, and

$$SS_Q = \frac{1}{pr} \sum_{j=1}^{q} Y^2_{.j.} - C$$

for the main effect of Q.

(iii) The interaction sum of squares is then given by subtraction as

$$SS_{PQ} = SS_T - SS_P - SS_Q.$$

We thus obtain the analysis of variance table

Source of Variation	Sum of Squares	Degrees of Freedom	Mean Square	F-Ratio
Between blocks	SS_B	$b - 1$	MS_B	$F_B = \frac{MS_B}{MS_E}$
Main effect P	SS_P	$p - 1$	MS_P	$F_P = \frac{MS_P}{MS_E}$
Main effect Q	SS_Q	$q - 1$	MS_Q	$F_Q = \frac{MS_Q}{MS_E}$
PQ interaction	SS_{PQ}	$(p - 1)(q - 1)$	MS_{PQ}	$F_{PQ} = \frac{MS_{PQ}}{MS_E}$
Error	SS_E	$n - pq - b + 1$	MS_E	
Total	S_{yy}	$n - 1$		

As always, the error mean square MS_E provides an estimate s_e^2 of σ^2 to be used in constructing confidence intervals for any factor means of particular interest. The ratios F_P, F_Q and F_I are referred to critical values of the $F_{p-1,n-pq-b+1}$, $F_{q-1,n-pq-b+1}$ and $F_{(p-1)(q-1),n-pq-b+1}$ distributions respectively to provide tests of significance for the two main effects and interaction.

In the absence of a significant interaction, significant main effects indicate different responses for different levels of the relevant factors, and valid summaries for each factor are provided by averaging the response over all other factors and blocks (if present). However, if the interaction is significant then care must be taken with interpretation and summary of results. In this case, the factors are *not independent*: the difference between responses to one factor will vary over levels of the other factor(s). To summarize, one therefore needs to give effects of each factor *for each level of the other factor(s)*. This can be done either graphically or in tables. Alternatively, these individual effects can be tested for significance by subdividing the main effect sum of squares. For example, to investigate the individual effects of factor P, we obtain $SS(P$ within level i of Q) for $i = 1, \ldots, q$ by applying the main effect sum of squares procedure to the totals at each separate level of Q. Each of these components has $p - 1$ degrees of freedom, so they can be tested for significance by computing their mean square ratios with MS_E in the usual way and comparing with the $F_{p-1, n-pq-b+1}$ distribution. However, they add up to $SS_P + SS_I$ (so that all possible comparisons obtained in this way will not be independent).

If there are no blocks in the experiment, and there is only a single observation at each combination of factor levels, then both SS_E and its degrees of freedom are zero and there is no estimate of σ^2. In this case it is impossible to conduct a full analysis that allows for interaction between the factors. The best that we can do is to assume additivity of the factors, in which case SS_I just becomes the error sum of squares and the rest of the analysis proceeds as before.

Example 4.4. Return to Example 4.2, i.e. to the additive randomized blocks analysis of tyre wear, but now let us take account of the factorial structure outlined above for the four groups: factor P ('direction') with levels 'front' and 'rear', and factor Q ('side') with levels 'off' and 'near'. Everything is exactly as before, except that group 1 is now a_1b_1, group 2 is a_1b_2, group 3 is a_2b_1 and group 4 is a_2b_2 (so that $p = q = 2$, $b = s = 3$ and $r = bs = 9$). Thus

Total of P at lower level $=$ sum of totals for groups 1 and 2 $= 352.255$,

Total of P at upper level $=$ sum of totals for groups 3 and 4 $= 542.635$,

Total of Q at lower level $=$ sum of totals for groups 1 and 3 $= 487.797$,

Total of Q at upper level $=$ sum of totals for groups 2 and 4 $= 407.093$.

Hence
$$C = 894.887^2/36 = 22245.225,$$

$$SS_P = \frac{1}{2 \times 9}[(352.255)^2 + (542.635)^2] - 22245.225 = 1006.793,$$

$$SS_Q = \frac{1}{2 \times 9}[(487.797)^2 + (407.093)^2] - 22245.225 = 180.921,$$

and

$$SS_I = 1189.014 - 1006.793 - 180.921 = 1.3$$

(since the sum of squares between all four groups was 1189.014 in Example 4.2). This yields the new ANOVA table:

Source	S.S.	d.f.	M.S.	F-Ratio
Between blocks	156.884	2	78.442	3.07
Main effect of P	1006.793	1	1006.793	39.4
Main effect of Q	180.921	1	180.921	7.08
PQ interaction	1.300	1	1.300	0.05
Error	766.235	30	25.541	
Total	2112.133	35		

Critical values of the $F_{1,30}$ distribution are 4.17 (for 5% significance), 7.56 (for 1% significance), and 13.29 (for 0.1% significance). Thus the main effect of P ('direction') is very highly significant, the main effect of Q ('side') is significant at about the 1% level, but the PQ interaction is not significant. Since the interaction is not significant, we are justified in summarizing the results by averaging one factor over the levels of the other, or by obtaining confidence intervals for the difference in means between levels of one factor without regard to the other. For example, if we want a 95% confidence interval for the difference in mean wear between all front and rear tyres, then from above we have the point estimate $542.635/18 - 352.255/18 = 10.5767$. Also $s_e = \sqrt{25.541} = 5.054$, so that the standard error for the difference between the two means is $s_e\sqrt{\frac{1}{18} + \frac{1}{18}} = 5.054 \div 3 = 1.685$, and for a 95% confidence interval the t-multiplier (30 degrees of freedom) is 2.04. The required interval is thus $10.5767 \pm 1.685 \times 2.04$, i.e. (7.14, 14.01).

Note that the sums of squares for main effect of P, main effect of Q, and PQ interaction derived above are precisely the same as the three single degree of freedom components of the between-groups sum of squares derived in Example 4.1. This is unsurprising, of course, as the sums of squares of factorial effects are seen from above to be orthogonal components of a between-groups sum of squares, and inspection of the particular components chosen in Example 4.1 shows them to correspond to the above main effects and interaction.

Presence of more than two factors

Experiments and surveys frequently involve more than two factors, and the above ideas all extend directly to three or more cross-classified factors. The main vehicle for analysis is once again analysis of variance, with the between-groups sum of squares being partitioned into components attributable to main effects and interactions (where the 'groups' are the separate combinations of the different factor levels).

Main effects are defined exactly as before, i.e. as comparisons among the responses at different levels of each factor averaged over all other factor combinations, but interactions cause a little more trouble. In particular, various *types* of interaction can now be identified, depending on how many factors are taken into account: *two-factor interactions, three-factor interactions*, and so on. If there is a BC interaction in a three-factor experiment with factors A, B and C, for example, then comparisons among responses to C will vary with the levels of B (or vice versa) *where responses are averaged over all levels of A*. An ABC interaction can then be viewed in several ways: as an interaction of the BC interaction with A (i.e. the BC interaction is different at different levels of A), or as an interaction of the AB interaction with C, or as an interaction of the AC interaction with B. The way to handle such an interaction will depened on which approach is most meaningful and upon the significance or otherwise of the two-factor interactions. Higher-order interactions are defined analogously. For example, an $ABCD$ interaction in a four-factor experiment with factors A, B, C and D can be viewed as an interaction of the ABC interaction with levels of D, and so on. However, interpretation of high-order interactions becomes very difficult in practice and the bulk of the emphasis in a typical analysis is generally placed on main effects and two-factor interactions only.

Partitioning of the between-group sum of squares proceeds in a fairly obvious fashion, by summing squares of totals in one-way tables (i.e. totals at levels of a single factor over all other factors) for main effects, those in two-way tables (i.e. totals at combinations of levels for two factors over all other factors) for two-factor interactions, and so on.

To illustrate, we give the details for a three-factor experiment with factors P, Q and S having p, q and s levels respectively. Moreover, we assume that there are also b blocks to take account of and that each of the pqs factor combinations is replicated r times (so that the total number of observations is $n = pqsr$). Let y_{ijku} denote the uth response at level i of P, level j of Q and level k of S, and denote totals as usual by upper-case letters: $Y_{i...}$ for the total of the qsr responses at level i of P, $Y_{ij..}$ for the total of the sr responses at level i of P and level j of Q, $Y_{ijk.}$ for the total of the r responses at level i of P, level j of Q and level k of S, and so on. We first obtain the standard analysis of variance, treating each of the pqs different factor combinations as a 'group'. Writing $C = Y^2_{....}/n$ for the correction factor we then have the following components of the between-group sum of squares SS_T:

$$SS_P = \frac{1}{qsr} \sum_i Y^2_{i...} - C; \qquad SS_Q = \frac{1}{psr} \sum_j Y^2_{.j..} - C$$

and

$$SS_S = \frac{1}{pqr} \sum_k Y^2_{..k.} - C$$

for the main effects;

$$SS_{PQ} = \frac{1}{sr} \sum_i \sum_j Y_{ij..}^2 - C - SS_P - SS_Q,$$

$$SS_{PS} = \frac{1}{qr} \sum_i \sum_k Y_{i.k.}^2 - C - SS_P - SS_S$$

and

$$SS_{QS} = \frac{1}{pr} \sum_j \sum_k Y_{.jk.}^2 - C - SS_Q - SS_S$$

for the two-factor interactions; and

$$SS_{PQS} = SS_T - SS_P - SS_Q - SS_S - SS_{PQ} - SS_{PS} - SS_{QS}$$

for the three-factor interaction. We thus obtain the analysis of variance table

Source of Variation	Sum of Squares	Degrees of Freedom
Between blocks	SS_B	$b - 1$
Main effect P	SS_P	$p - 1$
Main effect Q	SS_Q	$q - 1$
Main effect S	SS_S	$s - 1$
PQ Interaction	SS_{PQ}	$(p-1)(q-1)$
PS Interaction	SS_{PS}	$(p-1)(s-1)$
QS Interaction	SS_{QS}	$(q-1)(s-1)$
PQS Interaction	SS_{PQS}	$(p-1)(q-1)(s-1)$
Error	SS_E	$pqs(r-1) - b + 1$
Total	S_{yy}	$pqsr - 1$

Mean squares and F-ratios, leading to F-tests, are calculated as usual, so the relevant columns have been omitted from the table above.

Note again that when each treatment combination appears only once in the data (i.e. when $r = 1$ and $b = 1$) then the error sum of squares will be zero and we cannot estimate σ^2 from the full analysis above. In such cases we use the high-order interactions as 'error', so in a three-factor experiment we would use the three-factor interaction in this way. If there are more than three factors, it is common to pool the highest-order interactions into the error term, as these interactions are rarely interpretable in easy fashion.

Example 4.5. Let us return to Example 4.3, where we established that there was a significant groups by blocks interaction. We are now able to interpret and explain this interaction by treating the previous blocks (i.e. columns of the original Table 4.1) as levels of a third factor ('cars') to add to the two factors ('direction' and 'side') outlined in Example 4.4. In fact, the three cars were fundamentally different in type: car A (first column) was front-wheel drive with manual transmission (FM), car B (second column) was rear-wheel drive with manual transmission (RM), while car C was front-wheel drive with automatic transmission (FA). Hence we have a $3 \times 2 \times 2$ factorial arrangement with the following factors and levels:

Factor P ('direction') with levels F and R;

Factor Q ('side') with levels O and N;

Factor S ('cars') with levels FM, RM and FA.

There are $r = 3$ observations at each of the 12 combinations of factor levels, and the totals of these observations are as follows:

		Factor S		
Factor P	Factor Q	FM	RM	FA
F	O	60.280	48.327	85.986
F	N	58.868	38.248	60.546
R	O	86.606	116.097	90.501
R	N	58.177	105.354	85.900

The following quantities were obtained previously:

(i) Total sum of squares $S_{yy} = 2112.133$ (see Example 4.1) and correction factor $C = (894.887)^2/36 = 22245.225$ (see Example 4.4);

(ii) Between-combination sum of squares (treating the 12 factor combinations above as groups) $SS_C = 2084.549$ (see Example 4.3);

(iii) Error sum of squares $SS_E = 2112.133 - 2084.549 = 27.584$ (see Example 4.3).

To obtain the main effect and interaction breakdown of the sum of squares between combinations, we need to form all three two-way tables by summing the above totals in turn over the levels of each factor. (Note that the marginal totals thus occur twice, providing a check on the arithmetic).

First, summing over the levels of S we have the two-way table for P, Q:

	O	N	Total
F	194.593	157.659	352.252
R	293.204	249.431	542.635
Total	487.797	407.090	894.887

From this table we obtain:

(i) $SS_P = \frac{1}{3\times3\times2}[(352.252)^2 + (542.635)^2] - 22245.225 = 1006.793$,

(ii) $SS_Q = \frac{1}{3\times3\times2}[(487.797)^2 + (407.090)^2] - 22245.225 = 180.921$,

(iii) $SS_{PQ} = \frac{1}{3\times3}[(194.593)^2 + (157.659)^2 + (293.204)^2 + (249.431)^2] - 180.921 - 1006.793 - 22245.225 = 1.300$

(and we notice that these values agree with those in Example 4.4).

Next, summing over the levels of P we have the two-way table for Q, S:

	FM	RM	FA	Total
O	146.886	164.424	176.487	487.797
N	117.045	143.602	146.443	407.090
Total	263.931	308.026	322.930	894.887

From this table we obtain:

(i) $SS_S = \frac{1}{3\times2\times2}[(263.931)^2 + (308.026)^2 + (322.930)^2] - 22245.225 = 156.883$,

(ii) $SS_{QS} = \frac{1}{3\times2}[(146.886)^2 + \cdots + (146.443)^2] - 180.921 - 156.983 - 22245.225 = 4.622$

(since $SS_Q = 180.921$ from above).

Finally, $SS_{PS} = 638.278$ is obtained in exactly the same fashion from the two-way table for P, S. Then SS_{PQS} is obtained by subtracting all these derived quantities from SS_C:

$$2084.549 - 1006.793 - 180.921 - 1.300 - 156.883 - 4.622 - 638.278 = 95.873.$$

We are thus able to draw up the full ANOVA:

Source	S.S.	d.f.	M.S.	F-Ratio
Main effect of P	1006.793	1	1006.793	876
Main effect of Q	180.921	1	180.921	157
Main effect of S	156.883	2	78.442	68.3
PQ interaction	1.300	1	1.300	1.13
PS interaction	638.278	2	319.139	278
QS interaction	4.622	2	2.311	2.01
PQS interaction	95.873	2	47.876	41.7
Error	27.584	24	1.149	
Total	2112.133	35		

Looking up critical values in F-tables is unnecessary in this instance, as the calculated F-ratios are either enormous or very small. It is evident that neither the PQ nor the QS interactions are significant, but that all the other effects are very highly significant. In particular the PS and PQS interactions together account for the blocks by groups interaction that was detected in the simplified analysis of Example 4.3. To interpret the PS interaction, we need to examine the two-way table for P, S:

	FM	RM	FA
F	119.148	86.575	146.529
R	144.783	221.451	176.401

It is evident that, while tyre wear is greater on the rear wheels than on the front ones for all three types of car, the difference is about four times greater for the rear-wheel drive car than for the two front-wheel drive ones (but there is no distinction between manual and automatic transmission as regards this difference). This much larger difference in the middle column than in the two outer columns accounts for the PS interaction. In addition, presence of a PQS interaction tells us that the distinction between front-wheel and rear-wheel drive cars is not constant across level of the third factor ('side'). To interpret this aspect, we must examine the full three-way table of totals given at the start of the example. There we see that the difference between front and rear wheel wear is approximately the same for both offside and nearside tyres with the rear-wheel drive car, but not so for the two front-wheel drive cars. In the case of the manual transmission (first column) there is a bigger difference between rear and front for the offside than for the nearside (26.326 versus -0.691), but for the automatic transmission (third column) the reverse is the case (4.515 versus 25.357).

Thus a full factorial analysis of this set of data has given us a complete picture of all the influences acting on the responses. Note also how the gradual uncovering of extra information, and its incorporation into the model and analysis through Examples 4.1 to 4.5, has sharpened up the conclusions to be drawn from the experimental results. The practical lesson to be learnt here is that it is vital to obtain *all the relevant information* about an experiment or survey *before* undertaking any analysis of the results. While the conclusions drawn in Example 4.1 might be of interest to the experimenter, they are nowhere near as comprehensive as those in Example 4.5.

4.4 Nested classifications

4.4.1 Basic idea

Situations often arise in practice where a process of *subsampling* takes place before responses are measured on individuals. Consider the following examples.

(i) A clothing manufacturer wishing to investigate the properties of different chemical dyes on wool applies each dye to a large batch of material. From each large batch several small batches are chosen at random and some response, such as resistance to fading, is measured on each chosen batch.

(ii) A forester wishes to study the effects of three different sprays used on trees. He chooses 12 trees (from a large population of trees available to him) and sprays each chemical onto a randomly chosen four of these trees. After one week he picks six leaves at random from each tree and measures the concentration of nitrogen on each leaf.

(ii) A researcher wishes to measure the effect of government cuts on hospital beds in the National Health Service. She chooses four regional health authorities at random from all such authorities in the UK; within each authority she chooses three conurbations at random, within each conurbation she selects three hospitals at random, and within each hospital she selects three wards at random. She then records the number of beds lost over the previous month in each ward.

These are all examples of *nested* or *hierarchical* arrangements, in increasing *depth* of nesting. In (i) we have batches nested within dyes; in (ii) we have leaves nested within trees nested within sprays; and in (iii) we have wards nested within hospitals nested within conurbations nested within authorities. There are two fundamental differences between such arrangements and the cross-classifications discussed so far in this chapter, both stemming from the fact that sampling has been employed in the design of a nested investigation.

(a) The process of subsampling induces extra random variation over and above the usual experimental error. For example, there may be considerable between-hospital, or between-ward, variation in bed closures in (iii) and the actual responses recorded will depend crucially on which hospitals and wards have actually been chosen. A major aim of analysis for nested designs is thus to obtain estimates of such random variation at each level of nesting, and thereby to allow for them when drawing conclusions about the particular study.

(b) The process of subsampling removes the connections between the experimental units that exist in a cross-classification. For example, 'ward 1' in the first hospital chosen in (iii) has nothing in common with 'ward 1' in the second or third hospitals; 'leaf 1' in the first tree chosen in (ii) has no connection with 'leaf 1' in the other trees; and similarly for any of the units chosen at any level of any such design. Since they are all chosen purely randomly, and no interest attaches to the particular ones actually selected, they are said to exhibit *random effects*. By contrast, all the levels of factors, or blocks, in a cross-classification are common across all experimental units. Moreover, the actual levels used in the experiment are generally of interest in their own right, so are treated as having *fixed effects*.

These major aspects need to be dealt with in the model and analysis associated with a nested design. We now outline the essential features for a design with nesting to depth two; the same principles extend to any depth of nesting, and the elements of analysis extrapolate in the obvious way.

4.4.2 Model

The top level of the hierarchy often corresponds to different *treatments* applied to individuals [e.g. dyes in (i), sprays in (ii) above], so we will adopt this as standard nomenclature in the present section for convenience. We can therefore denote any particular response by y_{ijk}, where i is the number of the treatment $(i = 1, \ldots, t)$, j is the number of the experimental unit to which the ith treatment is applied $(j = 1, \ldots, b)$ and k is the number of the subsample from the jth unit receiving treatment i $(k = 1, \ldots, r)$. Then we can construct the following model:

$$y_{ijk} = \mu + \alpha_i + \beta_{(i)j} + \epsilon_{(ij)k}. \tag{4.16}$$

Here μ is the usual overall mean 'effect', and α_i is the effect of treatment i. Since the treatments are generally of interest in their own right and pre-selected, the α_i can be treated as fixed effects and therefore subject to the constraint $\sum_i \alpha_i = 0$.

$\beta_{(i)j}$ is then the effect of the jth unit receiving treatment i, and $\epsilon_{(ij)k}$ is the effect of the kth subunit from unit j receiving treatment i. The brackets in these subscripts denote nesting: β_{ij} in a cross-classification implies that when the same j is applied to different i we are referring to the *same* experimental unit, whereas here $\beta_{(i)j}$ implies that when the same j is applied to different i we are referring to *different* units. For the reasons outlined earlier, these nested parameters must be treated as random effects. The essential features of such effects are that they are zero on average but the actual value of any particular effect is a realization of a random variable. The simplest assumptions satisfying this requirement are thus that the $\beta_{(i)j}$ are iid $N(0, \sigma_b^2)$ random variables, the $\epsilon_{(ij)k}$ are iid $N(0, \sigma_e^2)$ random variables, and the β's and ϵ's are mutually independent.

4.4.3 Analysis

Analysis of variance

To obtain the ANOVA, we must divide the total sum of squares into meaningful parts. First, by adding and subtracting the terms $y_{i..}$ and $y_{ij.}$ and suitably bracketing the resulting expression we obtain the identity

$$y_{ijk} - y_{...} = (y_{i..} - y_{...}) + (y_{ij.} - y_{i..}) + (y_{ijk} - y_{ij.})$$

where, as usual, $y_{...}$ is the mean of all $n = tbr$ observations, $y_{i..}$ is the mean of the br observations receiving treatment i, and $y_{ij.}$ is the mean of the r observations in unit j that receive treatment i. When both sides of the identity are squared and added over all observations, all cross products vanish because they contain at least one sum of values about their mean and we obtain the sum of squares breakdown

$$S_{yy} = SS_T + SS_U + SS_E$$

where

$S_{yy} = \sum_i \sum_j \sum_k (y_{ijk} - y_{...})^2$ is, as usual, the total sum of squares,

$SS_T = br \sum_i (y_{i..} - y_{...})^2$ is the sum of squares between treatments,

$SS_U = r \sum_i \sum_j (y_{ij.} - y_{i..})^2$ is the sum of squares between units, pooled within treatments, and

$SS_E = \sum_i \sum_j \sum_k (y_{ijk} - y_{ij.})^2$ is the sum of squares between subunits, pooled within both treatments and units. (This is the sum of squares at the lowest level of the hierarchy, so is traditionally labelled the error sum of squares).

S_{yy} and SS_T have their usual number of degrees of freedom, viz. $n - 1$ and $t - 1$ respectively. SS_U is a sum of t separate sums of squares, each of which has $b - 1$ degrees of freedom, so has $t(b - 1)$ degrees of freedom. There are thus $tb(r - 1)$ degrees of freedom left over for SS_E. Hence we obtain the ANOVA table:

Source of Variation	Sum of Squares	Degrees of Freedom	Mean Square
Between treatments	SS_T	$t - 1$	$MS_T = SS_T/(t - 1)$
Between units within treatments	SS_U	$t(b - 1)$	$MS_U = SS_U/[t(b - 1)]$
Error	SS_E	$tb(r - 1)$	$MS_E = SS_E/[tb(r - 1)]$
Total	S_{yy}	$n - 1$	

Computation of the various sums of squares required in this ANOVA can be obtained most efficiently and conveniently as follows (using upper case as usual for totals).

(i) Correction factor: $C = Y_{...}^2/n$;

(ii) $S_{yy} = \sum_i \sum_j \sum_k y_{ijk}^2 - C$;

(iii) $SS_T = \frac{1}{br} \sum_i Y_{i..}^2 - C$;

(iv) $SS_U = \frac{1}{r} \sum_i \sum_j Y_{ij.}^2 - C - SS_T$;

(v) $SS_E = S_{yy} - SS_T - SS_U$.

Tests of hypothesis

This is the point at which the ANOVA procedure for a model involving random effects departs from that for models involving only fixed effects. All our previous situations in this chapter have been of the latter type, and testing for significance of any specific line in the ANOVA involved taking the ratio of that line's mean square to the error mean square and referring the result to the appropriate F-distribution. We now have to modify this procedure slightly.

First, it is relatively straightforward to establish that the expected values of the various mean squares in the above table are as follows:

(i) $E(MS_T) = \sigma_e^2 + r\sigma_b^2 + \frac{br}{t-1}\sum_i \alpha_i^2;$

(ii) $E(MS_U) = \sigma_e^2 + r\sigma_b^2;$

(iii) $E(MS_E) = \sigma_e^2.$

Thus we have the following consequences.

(a) If the null hypothesis of no treatment differences, i.e. $H_0 : \alpha_1 = \cdots = \alpha_t = 0$, is true then MS_T and MS_U are two independent estimates of $\sigma_e^2 + r\sigma_b^2$, but if the hypothesis is not true then MS_T should be 'bigger' than MS_U. A reasonable test statistic for this hypothesis is thus the ratio $F_T = MS_T/MS_U$, and it can be shown that this ratio has an F-distribution on $t - 1$ and $t(b - 1)$ degrees of freedom when H_0 is true. Hence the hypothesis is tested by referring F_T to this distribution, and is rejected at the $100\alpha\%$ level if the value of F_T exceeds the $100(1 - \alpha)\%$ point of this distribution.

(b) To test the hypothesis of no variability between units (i.e. all the variability in the experiment is due to subunits only), we would consider $H_0 : \sigma_b^2 = 0$. If this hypothesis is true then MS_U and MS_E are independent estimates of σ_e^2, but if the hypothesis is not true then MS_U should be 'bigger' than MS_E. By analogy with (a), therefore, we use the ratio $F_U = MS_U/MS_E$ as the test statistic and refer the calculated value to the $F_{t(b-1),tb(r-1)}$ distribution; large values of the ratio lead to rejection of the hypothesis.

Analogous results are obtained however many levels there are in the hierarchy of the nested structure, so that significance of any line in the ANOVA is tested by computing the ratio of that line's mean square to the mean square in the line *immediately below it* and referring the calculated value to the F-distribution with degrees of freedom given by those of the two lines. If the line relates to fixed effects (e.g. treatments), then the hypothesis tested is of no effects; but if the line relates to random effects then the hypothesis tested is of no variability at that level of the hierarchy.

Estimates and confidence intervals

As before, $y_{i..} - y_{...}$ is an unbiased estimate of the effect (α_i) of treatment i, and $y_{i..} - y_{j..}$ is an unbiased estimate of the difference in effects $(\alpha_i - \alpha_j)$ between treatments i and j. However, for confidence intervals we need the variances of these quantities and the variances are more complicated than before because of the presence of the random effects in the model. We find that the variance of $y_{i..} - y_{...}$ is $(t - 1)(\sigma_e^2 + r\sigma_b^2)/n$, and the variance of $y_{i..} - y_{j..}$ is

$2(\sigma_e^2 + r\sigma_b^2)/br$. From the expected mean squares above, we see that MS_U is an unbiased estimate of $\sigma_e^2 + r\sigma_b^2$, so we can use this estimate in the variances above to find standard errors of any desired treatment effect or difference of effects and hence to construct a confidence interval using the t-distribution.

For the random effects, MS_E is an unbiased estimate of σ_e^2 and MS_U is an unbiased estimate of $\sigma_e^2 + r\sigma_b^2$. Hence an unbiased estimate of σ_b^2 is given by $(MS_U - MS_E)/r$ if this is positive (and zero if it is not). Confidence intervals for σ_e^2 and σ_b^2 are obtained from the chi-squared distribution in standard fashion for variances (see Example 2.21).

Example 4.6. Steel and Torrie (*Principles and Procedures of Statistics, a Bio-metrical Approach*, McGraw Hill, 1980 p153) describe the following experiment into the effect of length of exposure to daylight on growth of plants. Six treatments (8, 12 and 16 hours exposure in each of two glasshouse conditions, low and high night temperatures), were assigned at random to pots with three pots per treatment, and four plants were taken at random from a large group and assigned to each pot. The growth in cm of the stems of the plants were measured after one week, and gave the following data (where treatments 1–3 are the three hours exposure for low night temperature and treatments 4–6 are the same three hours exposure at high night temperature):

Pot	Plant	Treatment 1	2	3	4	5	6
1	1	3.5	5.0	5.0	8.5	6.0	7.0
	2	4.0	5.5	4.5	6.0	5.5	9.0
	3	3.0	4.0	5.0	9.0	3.5	8.5
	4	4.5	3.5	4.5	8.5	7.0	8.5
2	1	2.5	3.5	5.5	6.5	6.0	6.0
	2	4.5	3.5	6.0	7.0	8.5	7.0
	3	5.5	3.0	5.0	8.0	4.5	7.0
	4	5.0	4.0	5.0	6.5	7.5	7.0
3	1	3.0	4.5	5.5	7.0	6.5	11.0
	2	3.0	4.0	4.5	7.0	6.5	7.0
	3	2.5	4.0	6.5	7.0	8.5	9.0
	4	3.0	5.0	5.5	7.0	7.5	8.0

Let y_{ijk} denote the week's growth for plant k in pot j receiving treatment i ($k = 1, \ldots, 4$; $j = 1, \ldots, 3$; $i = 1, \ldots, 6$). The treatment numbers are meaningful, but plant and pot numbers are simply for convenience in tagging the observations (e.g. plants numbered 1 have nothing in common but the number, pots numbered 2 have nothing in common but the number, and so on). This feature makes the arrangement in this experiment a nested one rather than a cross-classified one (in which pot 1, say, would have to be the *same* pot for all treatments).

Following the scheme given above for a nested design, therefore, we obtain the following values.

(i) Pot totals:

$$Y_{11.}=15.0, Y_{21.}=18.0, Y_{31.}=19.0, Y_{41.}=32.0, Y_{51.}=22.0, Y_{61.}=33.0$$

$$Y_{12.}=17.5, Y_{22.}=14.0, Y_{32.}=21.5, Y_{42.}=28.0, Y_{52.}=26.5, Y_{62.}=27.0$$

$$Y_{13.}=11.5, Y_{23.}=17.5, Y_{33.}=22.0, Y_{43.}=28.0, Y_{53.}=29.0, Y_{63.}=35.0$$

(ii) Treatment totals:

$$Y_{1..} = 44.0, Y_{2..} = 49.5, Y_{3..} = 62.5, Y_{4..} = 88.0, Y_{5..} = 77.5, Y_{6..} = 95.0$$

(iii) Grand total, $Y_{...} = 416.5$, Correction factor $C = 416.5^2/72 = 2409.34$.
(iv) $S_{yy} = 3.5^2 + 5.0^2 + \cdots + 7.5^2 + 8.0^2 - C = 255.91$.
(v) $SS_T = (44.0^2 + \cdots + 95.0^2)/12 - C = 179.64$.
(vi) $SS_U = (15.0^2 + 18.0^2 + \cdots + 29.0^2 + 35.0^2)/4 - C - SS_T = 25.83$.
(vii) $SS_E = 255.91 - 179.64 - 25.83 = 50.44$.

Thus we can draw up the analysis of variance table:

Source of Variation	Sum of Squares	Degrees of Freedom	Mean Square
Between treatments	179.64	5	35.93
Between pots within treatments	25.83	12	2.15
Error	50.44	54	0.93
Total	255.91	71	

A test of the null hypothesis of no treatment differences is thus given by the ratio $F_T = 35.93 \div 2.15 = 16.7$, and by reference to the $F_{5,12}$ distribution we see that this is significant at the 1% level. Thus we conclude that there *are* differences between the treatments. The standard error of a treatment mean is $\sqrt{2.15/12} = 0.42$, and the standard error of the difference between two means is $\sqrt{2(2.15)/12} = 0.60$, so differences between treatments can be explored using these standard errors.

The further test of no variability between pots is given by the ratio $F_U = 2.15/0.93 = 2.3$, and reference to the $F_{12,54}$ distribution shows this value to be (just) significant at the 5% level. Thus environmental differences between pots are in fact greater than variability within pots so do need to be taken account of in any discussion of results.

4.5   A general approach via multiple regression

We saw in Section 3.3.8 how categorical or qualitative explanatory variables could be incorporated into a multiple regression model by assigning numerical values to dummy quantitative variables for each level of every qualitative variable. In the present chapter we have been concerned exclusively with responses measured on different groups of individuals, where the grouping is determined by combinations of factors with or without the imposition of blocking of some form. We can view the factors and blocks as explanatory qualitative variables, so the methods of Section 3.3.8 can be employed to convert every model of the present chapter into a multiple regression model on dummy explanatory variables.

For example, consider model (4.2) in the case of a one-way arrangement. Here μ is present for all responses, so it is like the constant term β_0 in a multiple regression equation and can be written μX_0 where X_0 is a dummy variable taking values $x_{0ij} = 1$ for all i and j. We then have the k group effects α_i. The simplest way of incorporating these into a multiple regression model is as coefficients of k dummy variables $X_1, \ldots, X_k$ which indicate any individual response's group: if the response is for an individual in group t then $X_t = 1$ and $X_j = 0$ for $j \neq t$. However, if we do this then it is easy to see that we have a linear relationship among the dummy variables ($X_0 = X_1 + \cdots + X_k$) so the model is singular. We have to incorporate the constraint $\sum_i n_i \alpha_i = 0$ that is present in (4.2) in order to obtain an estimable model. The effect of this constraint is to reduce the number of parameters by one, because $\alpha_k = -\frac{n_1}{n_k}\alpha_1 - \cdots - \frac{n_{k-1}}{n_k}\alpha_{k-1}$. Thus we only retain $X_0, X_1, \ldots, X_{k-1}$ with the above definitions, and a response from group k is denoted by setting $X_i = -n_i/n_k$ for $i = 1, \ldots, k-1$. Model (4.2) is thus equivalent to the multiple regression model

$$Y = \mu X_0 + \alpha_1 X_1 + \cdots + \alpha_{k-1} X_{k-1} + \epsilon, \tag{4.17}$$

for these variables, and where the departure variable ϵ has elements as in (4.2).

In a similar way, the randomized block model in equation (4.7) can be expressed as the multiple regression model

$$Y = \mu X_0 + \alpha_1 X_1 + \cdots + \alpha_{k-1} X_{k-1} + \beta_1 X_k + \cdots + \beta_{b-1} X_{k+b-2} + \epsilon, \tag{4.18}$$

where the dummy explanatory variables are defined as follows:

- X_0, taking the value 1 for all responses, accounts for the overall mean as above;
- $X_1, \ldots, X_{k-1}$ are the group dummy variables such that if the response is for an individual in group i then $X_i = 1$ and $X_j = 0$ for $j \neq i$, unless $i = k$ in which case $X_j = -1$ for $j = 1, \ldots, k-1$;

- $X_k, \ldots, X_{k+b-2}$ are the block dummy variables such that if the response is for an individual in block i then $X_{k-1+i} = 1$ and $X_{k-1+j} = 0$ for $j \neq i$, unless $i = b$ in which case $X_{k-1+j} = -1$ for $j = 1, \ldots, b - 1$.

(Note that the model (4.7) is for an arrangement in which the number of observations is the same in each group, and the same in each block. This is the reason for the -1s rather than terms like $-n_i/n_k$ above.)

Any other model that forms the basis of a fixed-effects analysis of variance in the sense of the present chapter can be similarly expressed as a multiple regression model on dummy variables. The basic rules for coding these variables are as set out in Section 3.3.8. The main effect of a factor is given by an appropriate set of contrasts among the dummy variables associated with the levels of that factor (so that if the factor has two levels then there is just one dummy variable associated with it and its main effect is represented by this variable). Since an interaction is a departure from additivity, it is included as a set of *products* of appropriate dummy variables. Two-factor interactions involve products of pairs of dummy variables, three-factor interactions involve products of triples of dummy variables, and so on.

If we use the regression formulation of these models, and proceed with an analysis as in Chapter 3, we obtain the regression ANOVA breakdown of that chapter into a regression sum of squares and a residual sum of squares. The residual sum of squares in all cases equates to the error sum of squares of the present chapter (named 'within group' sum of squares in the one-way arrangement). The regression sum of squares is then the sum of the remaining sums of squares of the present chapter. Thus for the one-way arrangement it is just the between-groups sum of squares SS_T, for the randomized block design it is the sum of the between-blocks and between-groups sums of squares ($SS_B + SS_T$), and so on. To disentangle the separate elements of these sums of squares, we need to employ the methods of Section 3.4.1 for testing significance of specified subsets of variables. For example, to assess the significance of the groups in a randomized block design, we need to look at the sum of squares of the group parameters *allowing for the block parameters* in the terminology of that section. This sum of squares turns out to be equal to SS_T of the randomized block ANOVA in Section 4.3.1 above. Similarly, SS_B is equal to the sum of squares of the block parameters allowing for groups in the regression formulation. Moreover, many single degree-of-freedom contrasts can be formulated directly in terms of these parameters and hence the relevant sums of squares can be derived.

The reason for the parallel development of the methods in these two chapters is mainly historical. In the early days, before the advent of electronic computers, the calculations for a multiple regression analysis were very onerous. Consequently, wherever possible, alternative (simple) computational methods were developed. Such methods almost always involved some constraints on the allowable forms of data, and so it is in the present case. Although the computations in the present chapter are considerably simpler than those for a multiple regression, it should be evident that the situations dealt with are somewhat specialized and restricted. In general, ANOVAs along the lines of those in this chapter require *balanced* arrangements in which the same number of observations appear in each block, each factor and so on. When this is so then the corresponding regression models are at least partially orthogonal – for example, the block dummy variables are orthogonal to the group dummy variables in (4.18) – which simplifies matters considerably.

Thus if a set of data for analysis conforms to one of the arrangements of the present chapter, then the methods of this chapter are by far the easiest to use. However, once there are *any* departures from these ideal conditions (e.g. unequal numbers of observations in different blocks, lack of balance between different factors, etc) then the best way of ensuring a valid analysis is to set up a regression model on dummy variables and to use the methods of Chapter 3.

4.6   Analysis of covariance

Sometimes in a designed experiment, the response variable Y might be influenced by some other variable X that has been measured on each individual. For example, consider an animal feed trial on pigs to compare the effects of three food additives. The three additives form the three treatments for comparison, and to minimize heterogeneity among the experimental units (i.e. the individual pigs), litters are used as blocks. Blocking in this way will ensure that pigs within a litter are reasonably homogeneous in respect of their genetic features, but other factors may still introduce heterogeneity among them. In particular, if the response variable Y is the pig's weight at the end of the trial then the pig's weight X at the start of the trial might be expected to exert some influence. Moreover, there will be quite a range of initial weights in each block, so if we can allow for this variation in some way then we should increase the precision of the final results.

One way in which we might take account of the initial weight is by analysing the *weight gain* $Y - X$. However, this assumes that the effect of X is essentially constant and additive (i.e. that it doesn't matter what the actual value of X is when adjusting for its effect). A more realistic assumption is that there is a general (but unknown) *linear* effect of X on the outcome Y, and such an assumption can be included in the model on which the analysis is based. Thus if we assume no block × treatment interaction, then an appropriate model would be

$$y_{iju} = \mu + \alpha_i + \beta_j + \gamma(x_{iju} - \overline{x}) + \epsilon_{iju} \qquad (4.19)$$

where y_{iju}, x_{iju} are the final and initial weights respectively for the uth pig in the jth litter given the ith treatment, μ, α_i and β_j are the overall mean and the usual treatment and block effects as specified in equation (4.7), and the ϵ_{iju} are iid $N(0, \sigma^2)$. The regression parameter γ indicates the strength of effect of X on Y.

Such a variable X is known as a *concomitant* variable, or *covariate*, and incorporating it into the analysis leads to an *analysis of covariance*. The one important assumption implicit in the above model is that *the relationship between x and y is unaffected by either blocks or treatments*. This assumption is reflected by the fact that γ does not vary with either i or j, but takes a single value across the experiment. Such an assumption makes sense, because otherwise the regression effect would become intertwined with either the block or the treatment effects (or both), and it would be difficult to disentangle all these different effects.

Given model (4.18), the easiest mode of analysis is the general multiple regression approach outlined in the previous section. We simply incorporate the covariate as an extra term in the equation, and then look at treatment and block sums of squares 'allowing for the covariate': using the methods of Section 3.4.1, this means looking at the difference between the model that includes all terms and the model that includes the covariate only. However, in the same way that special analysis of variance formulations have been set out earlier in the chapter for a number of standard designs, so it is relatively straightforward to derive analysis of *covariance* formulations for the corresponding designs. We will sketch out briefly the relevant computations for the additive block design given above; corresponding calculations for other designs can then be deduced in obvious fashion. As before, assume that we have k treatments applied to n experimental units in such a way that there are b blocks and r units receive each treatment in each block.

First, if we rewrite equation (4.19) in the following way

$$y_{iju} - \gamma(x_{iju} - \bar{x}) = \mu + \alpha_i + \beta_j + \epsilon_{iju}, \qquad (4.20)$$

it should be evident that the analysis consists of a standard randomized block analysis on the values of Y after adjustment for the regression of Y on X. Thus one straightforward approach is a two-stage analysis: carry out a regression of Y on X, then conduct a randomized block analysis on the residuals from this regression.

However, these two stages can be rolled together quite easily. Recollect from Section 4.3.1 the randomized block ANOVA:

Source of Variation	Sum of Squares	Degrees of Freedom	Mean Square	F-Ratio
Between blocks	B_{yy}	$b - 1$	MS_B	$F_B = \frac{MS_B}{MS_E}$
Between treatments	T_{yy}	$k - 1$	MS_T	$F_T = \frac{MS_T}{MS_E}$
Error	E_{yy}	$n - k - b + 1$	MS_E	
Total	S_{yy}	$n - 1$		

where we have slightly changed the notation for the sums of squares, but calculate them in the usual way:

$$S_{yy} = \sum_{i=1}^{k}\sum_{j=1}^{b}\sum_{u=1}^{r} y_{iju}^2 - ny_{...}^2,$$

$$T_{yy} = \sum_{i=1}^{k} (Y_{i..}^2/br) - Y_{...}^2/n,$$

$$B_{yy} = \sum_{j=1}^{b} (Y_{.j.}^2/kr) - Y_{...}^2/n,$$

and E_{yy} is obtained by subtraction $(S_{yy} - T_{yy} - B_{yy})$.

To obtain an analysis of covariance, we then apply the same formulae (carrying over the notation in an obvious fashion) successively to the sums of squares of xs:

$$S_{xx} = \sum_{i=1}^{k}\sum_{j=1}^{b}\sum_{u=1}^{r} x_{iju}^2 - nx_{...}^2,$$

$$T_{xx} = \sum_{i=1}^{k} (X_{i..}^2/br) - X_{...}^2/n,$$

$$B_{xx} = \sum_{j=1}^{b} (X_{.j.}^2/kr) - X_{...}^2/n,$$

$$E_{xx} = S_{xx} - T_{xx} - B_{xx},$$

and to the sums of products of xs and ys:

$$S_{xy} = \sum_{i=1}^{k}\sum_{j=1}^{b}\sum_{u=1}^{r} x_{iju}y_{iju} - nx_{...}y_{...},$$

$$T_{xy} = \sum_{i=1}^{k} (X_{i..}Y_{i..}/br) - X_{...}Y_{...}/n,$$

$$B_{xy} = \sum_{j=1}^{b} (X_{.j.}Y_{.j.}/kr) - X_{...}Y_{...}/n,$$

$$E_{xy} = S_{xy} - T_{xy} - B_{xy},$$

and write all these quantities formally in an analysis of covariance (ANCOVA) table:

Source of Variation	$SS(y)$	$SP(xy)$	$SS(x)$	DF
Blocks	B_{yy}	B_{xy}	B_{xx}	$b-1$
Treatments	T_{yy}	T_{xy}	T_{xx}	$k-1$
Error	E_{yy}	E_{xy}	E_{xx}	$n-k-b+1$
Total	S_{yy}	S_{xy}	S_{xx}	$n-1$

We are now in a position to conduct a full analysis.

(i) From standard regression theory (Sections 3.2.2 and 3.2.3) we obtain the estimated regression coefficient

$$\hat{\gamma} = \frac{E_{xy}}{E_{xx}},$$

the regression sum of squares

$$RSS = \frac{E_{xy}^2}{E_{xx}}$$

on one degree of freedom, and the residual sum of squares

$$ESS = E_{yy} - \frac{E_{xy}^2}{E_{xx}}$$

on $n - k - b$ degrees of freedom. Comparing the ratio of regression and residual mean squares with the F-distribution on 1 and $n - k - b$ degrees of freedom tests the significance of relationship between Y and X, and the residual mean square gives the estimated variance s^2 of observations *after eliminating variation due to X*. We can therefore term this the adjusted residual mean square.

(ii) Making a similar adjustment for X to the treatments leads to slightly lengthier algebra but results in the adjusted treatment sum of squares

$$(T_{yy} + E_{yy}) - \frac{(T_{xy} + E_{xy})^2}{(T_{xx} + E_{xx})} - ESS$$

on $k - 1$ degrees of freedom. The ratio of adjusted treatment mean square and adjusted residual mean square can then be compared with the F-distribution on $k-1$ and $n-k-b$ degrees of freedom to test for significance of treatments *after eliminating variation due to X*.

(iii) The ith treatment mean adjusted for variation due to X is

$$y_{i..} - \hat{\gamma}(x_{i..} - x_{...}),$$

the difference between the ith and jth treatment means adjusted for variation due to X is

$$(y_{i..} - y_{j..}) - \hat{\gamma}(x_{i..} - x_{j..}),$$

and the standard error of this difference is

$$\sqrt{\left\{ s^2 \left[\frac{2}{br} + \frac{(x_{i..} - x_{j..})^2}{E_{xx}} \right] \right\}}.$$

This enables pairwise tests of treatments to be conducted, and confidence intervals to be calculated.

The general structure illustrated above for adjusting for a covariate in the randomized block design holds good for all other balanced designs for which a standard ANOVA exists: we apply the relevant formulae to obtain sums of squares of xs and sums of products of xs and ys as well as sums of squares of ys, and then adjust effects as above.

Example 4.7. Mead and Curnow (1983) give the following analysis of an experiment in which four different hormone treatments (A, B, C, D) were compared on sixteen steers grouped into four blocks. The response Y was the weight of kidney fat in grams measured on the carcasses, and the covariate X was the initial weight of the steers. X and Y were measured to the nearest 10 units (kilograms and grams respectively), and the data coded in units of 10 were as follows.

Block	A		B		C		D		Totals	
	x	y	x	y	x	y	x	y	x	y
I	56	133	44	128	53	129	69	134	222	524
II	47	132	44	127	51	130	42	125	184	514
III	41	127	36	127	38	124	43	126	158	504
IV	50	132	46	128	50	129	54	131	200	520
Totals	194	524	170	510	192	512	208	516	764	2062

We thus have $k = b = 4$, $r = 1$ and $n = 16$ in the notation above. Following the computational formulae above, we arrive at the ANCOVA summary:

Source of Variation	$SS(y)$	$SP(xy)$	$SS(x)$	DF
Blocks	57	174	545	3
Treatments	29	37	185	3
Error	42	85	243	9
Total	128	296	973	15

Without any adjustment for initial weight of steers, we have a hormone mean square of $29 \div 3 = 9.67$, an error mean square of $42 \div 9 = 4.67$ and an F-ratio for significance of hormone effects of $9.67 \div 4.67 = 2.07$, which is not significant.

After adjusting for initial weight, however, we have a residual sum of squares of $42 - 85^2/243 = 12.2675$ on 8 degrees of freedom, which gives a mean square of 1.53. This is much smaller than the unadjusted estimate of σ^2, indicating that covariance adjustment here increases the precision appreciably. From the formula above we find the adjusted hormone SS to be 24 on 3 degrees of freedom. The hormone mean square is thus 8, so the F-ratio for significance of hormones is $8/1.53 = 5.23$ which is now significant at the 5% level. Adjusting for initial weight has thus led to detection of a difference between the hormones.

Given this difference, we can now calculate the adjusted hormone means. The regression coefficient is $\hat{\gamma} = 85/243 = 0.350$, from which we obtain 130.7, 129.3, 127.9 and 127.5 as the adjusted means of hormones A, B, C and D respectively. Standard errors of differences between adjusted means depend on the treatments chosen, but range from 0.91 (for A–D) to 1.15 (for B–D), with an average of 0.98. It thus seems to be the difference between A and the others that produces the significant effect.

5
Non-normality: the theory of generalized linear models

5.1 Introduction

All the methods of Chapters 3 and 4 have been built round the general linear model (3.10), and therefore depend critically on the assumptions inherent in this model. Although we have seen that a large number of different situations can be handled by the model, it is nevertheless still a fairly restrictive one in some ways and hence would benefit from further generalization. In this chapter we outline a generalization that was first proposed in the 1970s and that now affords great flexibility in modelling all sorts of data types and situations. The general linear model (3.10) itself is a special case, as are various individual and *ad hoc* models used in earlier years for analysis of particular situations in which (3.10) is not appropriate. However, the great strength of this generalization is that it now provides a unified structure for all these types of situations, and enables the user to formulate suitable models and derive appropriate analyses not only for them but also for very specialized situations that are not specifically covered in text books.

In this chapter we consider all the *general* theory associated with the model: its definition, methods of estimating its parameters and testing hypotheses about them, ways of measuring the fit of the model and inspecting it for anomalies, and methodology for choosing parsimonious models. In succeeding chapters we consider the most common *particular* applications of the model, and develop techniques for analysing the data in these cases. Thus the present chapter is similar in tenor to Chapter 2, in that it develops general statistical ideas in a wide setting, while the next three chapters will provide analogous techniques to those of Chapters 3 and 4 for data that cannot be handled by the methods of those chapters.

5.2 The generalized linear model

To motivate and introduce the generalization, let us restate the general linear model (3.10), as it has been applied in Chapters 3 and 4, in the following way. Each of the *n* individuals in a data set has values observed for a *dependent* variable Y and p *explanatory* variables $X_1, \ldots, X_p$. Denote the value of the dependent variable for the ith individual by y_i and the explanatory variable values for the same individual by $x_{i1}, x_{i2}, \ldots, x_{ip}$. Since the latter values will in general influence the parameters of the distribution from which the dependent variable values come, we need to associate each y_i with its own random variable Y_i for full generality. The general linear model then has the following features.

(i) The Y_i are mutually independent normal random variables, the ith having mean μ_i and constant variance σ^2, i.e. $Y_i \sim N(\mu_i, \sigma^2)$ for $i = 1, \ldots, n$.

(ii) The explanatory variables provide a set of linear predictors $\eta_i = \beta_1 x_{i1} + \beta_2 x_{i2} + \cdots + \beta_p x_{ip}$ for $i = 1, \ldots, n$ (where, if necessary, the first explanatory variable is a dummy variable whose values all equal 1 to give a constant term in the model).

(iii) The link between (i) and (ii) is that $\mu_i = \eta_i$, i.e. the mean of the dependent variable for any observation is the linear predictor formed from that observation's values on the explanatory variables.

There are two main ways in which this model may be unsatisfactory in a given practical situation:

(a) the distribution of the dependent variable may not be normal;

(b) the mean of the dependent variable may be a *function* of the linear predictor, rather than just the linear predictor itself.

To illustrate, consider the following example.

Example 5.1. A toxicologist is investigating the efficacy of a new insecticide, and applies varying doses of the toxin to groups of selected insects. Let the dose (usually measured in logarithmic units) be denoted by X, and suppose that there are k groups of insects, with n_i in group i ($1 = 1, \ldots, k$). The ith group of insects is given a dose x_i of the toxin, and y_i of the insects in this group are still alive after a fixed period of time t. We are interested in drawing inferences about the probability π_x that a random insect will survive for time t after being given a dose $X = x$ of the toxin.

The dependent variable here can either be Y, the number of insects surviving, or $Z = Y/N$, the proportion of insects surviving a given dose of toxin. However, fitting a linear model such as

$$z_i (= y_i/n_i) = \alpha + \beta x_i + \epsilon_i \quad \text{with} \quad \epsilon_i \quad \text{iid} \quad N(0, \sigma^2) \qquad \text{for } i = 1, \ldots, k$$

would be inappropriate for a number of reasons. First, Y is not a normal variable but a binomial variable (a consequence of which is that the variance of Y will not be constant, either). Second, plotting the proportion of insects surviving in each group against the dose administered to that group does not generally yield a straight line, but usually exhibits a curved shape (typically sigmoidal, because mortality tends to rise more steeply in the centre of the dose range than at the extremes where either the dose is too low below one threshold to have much effect or so high above a second threshold that most insects are killed automatically). Finally, fitting a straight line regression as implied by the model above would often lead to predicted values of Z lying outside their permitted range $(0, 1)$.

A much more appropriate model to fit is one in which the distribution of y_i is correctly assumed to be binomial with parameters n_i and π_i for $i = 1, \ldots, k$ (where π_i denotes π_x at $x = x_i$), one which does not produce predicted values of Z_i outside the range $(0, 1)$, and one which allows for the sigmoidal pattern of proportion surviving. One way of achieving both of the latter objectives is to set $\pi_i = \Phi(\alpha + \beta x_i)$ where $\Phi(\cdot)$ is the cumulative normal distribution function (see Section 2.3.2). Thus, since the mean μ of the binomial distribution with parameters n and π is $n\pi$, we have the link $\mu = n\Phi(\eta)$ between the mean of Y and the linear predictor $\eta = \alpha + \beta X$ in this case. Moreover, since $\Phi(\cdot)$ only takes values between 0 and 1, values of Z predicted from this model can never lie outside their permitted range.

Motivated by such examples, and to overcome objections (a) and (b) above, Nelder and Wedderburn introduced in 1972 the class of *generalized linear models*. These models overcome objection (a) by being applicable whenever the dependent variable comes from any distribution in a wide class of distributions known as the *exponential family*. This family includes most of the distributions commonly encountered in practical situations, so the models are very widely applicable. Objection (b) is then overcome directly by allowing a general function to link the dependent variable mean and the linear predictor. We first consider details of the permitted distributions, and then describe the class of models.

5.2.1 The exponential family

A random variable Y has a distribution within the *exponential family* if its probability density (or mass) function $f(y)$ can be written in the *canonical form*

$$f(y; \theta, \phi) = \exp\{[y\theta - b(\theta)]/a(\phi) + c(y, \phi)\} \tag{5.1}$$

for some specific functions $a(\cdot)$, $b(\cdot)$ and $c(\cdot)$ and parameters θ and ϕ. Also, the range R of definition of Y must not involve either of these parameters.

If ϕ is known then the family is termed the *linear* exponential family and θ is the *natural*, or *canonical*, parameter. Quite often $a(\phi)$ is either of the form $a\phi$ for some constant a, or just ϕ where ϕ is σ^2 or 1. In such circumstances ϕ is termed the *scale* or *dispersion* parameter, and this is the case in most examples below.

Since $f(y; \theta, \phi)$ is a density, then

$$\int_R \exp\{[y\theta - b(\theta)]/a(\phi) + c(y, \phi)\}\, dy = 1.$$

Differentiating both sides of this equation with respect to θ yields

$$\int_R \frac{[y - b'(\theta)]}{a(\phi)} \exp\{[y\theta - b(\theta)]/a(\phi) + c(y, \phi)\}\, dy = 0$$

where $b'(\theta) = \frac{db(\theta)}{d\theta}$, the absence of θ from R enabling the differentiation on the left-hand side to be taken under the integral. Multiplying through by $a(\phi)$ and rearranging terms we obtain

$$\int_R y \exp\{[y\theta - b(\theta)]/a(\phi) + c(y, \phi)\}\, dy =$$

$$b'(\theta) \int_R \exp\{[y\theta - b(\theta)]/a(\phi) + c(y, \phi)\}\, dy.$$

Hence it follows that

$$E(Y) = b'(\theta). \tag{5.2}$$

Differentiating a second time with respect to θ yields

$$-\frac{b''(\theta)}{a(\phi)} \int_R \exp\{[y\theta - b(\theta)]/a(\phi) + c(y, \phi)\}\, dy +$$

$$\int_R \frac{[y - b'(\theta)]^2}{a^2(\phi)} \exp\{[y\theta - b(\theta)]/a(\phi) + c(y, \phi)\}\, dy = 0,$$

where $b''(\theta) = \frac{d^2 b(\theta)}{d\theta^2}$. In view of (5.2), this gives

$$\frac{1}{a^2(\phi)} \operatorname{Var}(Y) = \frac{b''(\theta)}{a(\phi)},$$

i.e.

$$\operatorname{Var}(Y) = a(\phi) b''(\theta). \tag{5.3}$$

With one eye on the specification of generalized linear models that is given next, we note that $E(Y)$ is often denoted by μ. It thus follows that $b'(\theta) = \mu$. Moreover, since $b''(\theta)$ depends on μ via $b'(\theta)$ it can be written as $V(\mu)$; this latter is often called the *variance function* of the model.

Most of the common distributions that are listed in Chapter 2 and used frequently in practice are members of the exponential family. Some examples are as follows.

Example 5.2. A normal variable Y with mean μ and variance σ^2 has density

$$f(y) = \frac{1}{\sqrt{(2\pi\sigma^2)}} \exp\{-(y - \mu)^2/2\sigma^2\}$$

$$= \exp\{(y\mu - \mu^2/2)/\sigma^2 - \frac{1}{2}[y^2/\sigma^2 + \log(2\pi\sigma^2)]\},$$

which is of the form (5.1) for $\theta = \mu$, $\phi = \sigma^2$, $a(\phi) = \phi$, $b(\theta) = \theta^2/2$, and $c(y, \phi) = -\frac{1}{2}[y^2/\phi + \log(2\pi\phi)]$. Hence $b'(\theta) = \theta = \mu$ and $b''(\theta) = 1$, so that (5.2) and (5.3) give $E(Y) = \mu$ and $\operatorname{Var}(Y) = \phi = \sigma^2$ as required.

Example 5.3. A binomial variable Y with parameters n (number of trials) and π (probability of success in each trial) has mean $\mu = n\pi$ and variance $\sigma^2 = n\pi(1 - \pi)$. The density is

$$f(y) = \binom{n}{y} \pi^y (1 - \pi)^{n-y}$$

$$= \binom{n}{y} (\pi/[1 - \pi])^y (1 - \pi)^n$$

$$= \exp\left\{ y \log(\pi/[1 - \pi]) + n \log(1 - \pi) + \log \binom{n}{y} \right\},$$

which is of the form (5.1) with $\theta = \log(\pi/[1 - \pi])$ and hence $\pi = e^\theta/(1 + e^\theta)$, $a(\phi) = 1$ (so that we can take $a(\phi) = \phi$ for $\phi = 1$), $b(\theta) = n \log(1 + e^\theta)$ and $c(y, \phi) = \log \binom{n}{y}$. Hence $b'(\theta) = \frac{ne^\theta}{1+e^\theta} = n\pi$ and $b''(\theta) = \frac{ne^\theta}{(1+e^\theta)^2} = n\pi(1-\pi)$, so that (5.2) and (5.3) give $E(Y) = n\pi$ and $\text{Var}(Y) = n\pi(1 - \pi)$ as required.

Example 5.4. A Poisson variable Y with parameter λ has mean $\mu = \lambda$ and variance $\sigma^2 = \lambda$. The density is

$$f(y) = \frac{\lambda^y}{y!} e^{-\lambda}$$

$$= \exp\{y \log \lambda - \lambda - \log y!\},$$

which is of the form (5.1) with $\theta = \log \lambda$, $a(\phi) = 1$ (so that we can again take $a(\phi) = \phi$ for $\phi = 1$), $b(\theta) = e^\theta$ and $c(y, \phi) = - \log y!$. Hence $b'(\theta) = e^\theta = \lambda$ and $b''(\theta) = e^\theta = \lambda$, so that (5.2) and (5.3) give $E(Y) = \lambda$ and $\text{Var}(Y) = \lambda$ as required.

5.2.2   The model

We are thus now in a position to define formally the generalized linear model. For the same situation as that in the specification of the general linear model at the start of this section, and using the same notation, the generalized linear model has the following features.

(i) The Y_i ($i = 1, \ldots, n$) are independent random variables sharing the same form of distribution from the exponential family.

(ii) The explanatory variables provide a set of linear predictors $\eta_i = \beta_1 x_{i1} + \beta_2 x_{i2} + \cdots + \beta_p x_{ip}$ for $i = 1, \ldots, n$.

(iii) The link between (i) and (ii) is that $g(\mu_i) = \eta_i$, where μ_i is the mean of Y_i for $i = 1, \ldots, n$; $g(\cdot)$ is called the *link function* of the model.

The two extensions to the general linear model that characterize the generalized linear model are thus its applicability to *any* member of the exponential family of distributions, and the presence of a link *function* when connecting the linear predictor η to the mean μ of Y. In any given problem there may of course be several feasible link functions that could be used, and the analyst may have to try each one before deciding on the best model to use. However, a particular simplification is introduced if the chosen link function is the same as the function that defines the canonical parameter for the relevant distribution; this link function is termed the *canonical link*. The canonical links for some standard distributions are as follows:

- binomial distribution, logit link $g(\mu) = \log\{\mu/(n - \mu)\}$;
- Poisson distribution, log link $g(\mu) = \log \mu$;
- normal distribution, identity link $g(\mu) = \mu$;
- gamma distribution, reciprocal link $g(\mu) = 1/\mu$;
- inverse normal, inverse square link $g(\mu) = 1/\mu^2$,

so these are the most commonly used link functions.

Example 5.5. The model suggested in the second part of Example 5.1 satisfies the requirements of a generalized linear model. First, the binomial distribution has been shown in Example 5.3 to be a member of the exponential family. Second, the linear predictor is $\alpha + \beta x_i$, where x_i is the dose of toxin applied to the ith group of insects. Third, the model specified for the binomial parameter is $\pi_i = \Phi(\alpha + \beta x_i)$. Now if Y_i is the number of insects surviving the ith dose then the mean of Y_i from the binomial distribution is $\mu_i = n_i \pi_i$. Hence this model can be re-expressed as $\mu_i = n_i \Phi(\alpha + \beta x_i)$, so that the linear predictor $\alpha + \beta x_i$ is equal to $\Phi^{-1}(\mu_i/n_i)$. Thus the link function is given by $g(\mu_i) = \Phi^{-1}(\mu_i/n_i)$.

Note that instead of modelling the sigmoidal relationship between proportion surviving and dose by the cumulative *normal* distribution function, we could have used the cumulative distribution function of any other suitable random variable instead. Choosing a logistic distribution, we would obtain the link function

$$g(\mu_i) = \log\left(\frac{\mu_i}{n_i - \mu_i}\right).$$

Re-expressing this function in terms of π_i instead of μ_i, we have

$$g(\pi_i) = \log\left(\frac{\pi_i}{1 - \pi_i}\right).$$

In showing that the binomial distribution was a member of the exponential family in Example 5.3, we found that the canonical parameter was given by $\theta = \log(\pi/[1 - \pi])$, which establishes that the link function obtained from the logistic distribution is in fact the canonical link. It is known as the *logit* link, and is the most popular link function for such data.

5.3 Fitting the model

Suppose we have a set of n random variables $Y_1, Y_2, \ldots, Y_n$ that satisfy the conditions (i) to (iii) of a generalized linear model as given in Section 5.2.2. The parameter θ depends on the explanatory variables $X_1, X_2, \ldots, X_p$ so can take different values for different Ys. Also, we assume ϕ to be constant across the Ys, but permit the function $a(\cdot)$ to vary. Thus we write θ_i and $a_i(\phi)$ for these quantities when applied to Y_i $(i = 1, \ldots, n)$. From (5.1) it then follows that the likelihood is given by

$$L(\theta_1, \ldots, \theta_n; \phi) = \prod_{i=1}^{n} \exp\{[y_i\theta_i - b(\theta_i)]/a_i(\phi) + c(y_i, \phi)\}, \qquad (5.4)$$

so that the log-likelihood is

$$l(\theta_1, \ldots, \theta_n; \phi) = \sum_{i=1}^{n} \frac{[y_i\theta_i - b(\theta_i)]}{a_i(\phi)} + \sum_{i=1}^{n} c(y_i, \phi). \qquad (5.5)$$

Of course the quantities of real interest are the parameters $\beta_1, \beta_2, \ldots, \beta_p$ of the linear predictor, and to obtain maximum likelihood estimates of these quantities we must solve the system of simultaneous equations

$$\frac{\partial}{\partial\beta_j} l(\theta_1, \ldots, \theta_n; \phi) = \sum_{i=1}^{n} \frac{\partial l_i}{\partial\beta_j} = 0 \qquad \text{for } j = 1, 2, \ldots, p \qquad (5.6)$$

where

$$l_i = \frac{[y_i\theta_i - b(\theta_i)]}{a_i(\phi)} + c(y_i, \phi).$$

By the chain rule of differential calculus we can write

$$\frac{\partial l_i}{\partial\beta_j} = \frac{\partial l_i}{\partial\theta_i} \times \frac{\partial\theta_i}{\partial\mu_i} \times \frac{\partial\mu_i}{\partial\eta_i} \times \frac{\partial\eta_i}{\partial\beta_j}. \qquad (5.7)$$

From the definition of l_i above we obtain directly

$$\frac{\partial l_i}{\partial\theta_i} = \frac{y_i - b'(\theta_i)}{a_i(\phi)}.$$

Moreover, since $\mu_i = b'(\theta_i)$ it follows that

$$\frac{\partial l_i}{\partial\theta_i} = \frac{(y_i - \mu_i)}{a_i(\phi)}$$

and

$$\frac{\partial\mu_i}{\partial\theta_i} = b''(\theta_i) = V(\mu_i).$$

Finally, the linear predictor $\eta_i = g(\mu_i) = \beta_1 x_{i1} + \beta_2 x_{i2} + \cdots + \beta_p x_{ip}$ supplies the last two terms required:

$$\frac{\partial \eta_i}{\partial \beta_j} = x_{ij}$$

and

$$\frac{\partial \eta_i}{\partial \mu_i} = g'(\mu_i) \quad \text{so} \quad \frac{\partial \mu_i}{\partial \eta_i} = \frac{1}{g'(\mu_i)}.$$

Substituting all these expressions into (5.7), we see from (5.6) that the maximum likelihood estimators of the β_j are obtained by solving the simultaneous equations

$$\sum_{i=1}^{n} \frac{(y_i - \mu_i)x_{ij}}{a_i(\phi)V(\mu_i)g'(\mu_i)} = 0 \qquad \text{for } j = 1, 2, \ldots, p. \tag{5.8}$$

These equations will of course depend on the particular generalized linear model being fitted, but in general they require numerical iterative techniques for their solution. Such techniques take an initial estimate β_0, say, of $\beta = (\beta_1, \ldots, \beta_p)'$ and then successively improve it to give a sequence of estimates $\beta_1, \beta_2, \beta_3, \ldots$ until the difference between successive vectors is less than some preset tolerance. The process is then said to have *converged*, and the resulting vector is the maximum likelihood estimator $\hat{\beta}$ of β.

There are two common iterative schemes in practice, the *Newton–Raphson* method and the *Method of Scoring*, respectively. In the Newton–Raphson procedure the rth iteration is given by

$$\beta_r = \beta_{r-1} - H_{r-1}^{-1} u_{r-1},$$

where u_{r-1} denotes the vector of first derivatives

$$\left(\frac{\partial l}{\partial \beta_1}, \frac{\partial l}{\partial \beta_2}, \ldots, \frac{\partial l}{\partial \beta_p} \right)'$$

evaluated at $\beta = \beta_{r-1}$ and H_{r-1} is the matrix of second derivatives that has (j, k)th element

$$\left(\frac{\partial^2 l}{\partial \beta_j \partial \beta_k} \right),$$

also evaluated at $\beta = \beta_{r-1}$.

The Method of Scoring is the same, except that the matrix H_{r-1} is replaced by the matrix I_{r-1} of *expected* second derivatives, with (j, k)th element

$$E\left(\frac{\partial^2 l}{\partial \beta_j \partial \beta_k} \right)$$

evaluated at $\beta = \beta_{r-1}$. A little bit of algebra establishes that this element is equal to

$$\sum_{i=1}^{n} \frac{x_{ij} x_{ik} (\partial \mu_i / \partial \eta_i)^2}{\mathrm{Var}(Y_i)},$$

evaluated of course at $\beta = \beta_{r-1}$.

A suitable starting vector β_0 for either of these schemes can be derived by regressing $g(Y)$ on the explanatory variables (i.e. by taking $\mu_i = y_i$ for all i in the linear predictor). Furthermore, it should be noted that when $g(\mu_i)$ is the canonical link then the matrix of second derivatives does not depend on the observed y_i. In this case H and I are the same, so the Newton–Raphson method is the same as the Method of Scoring.

Moreover, it is easily shown that the Method of Scoring iterative scheme can be expressed very concisely in the matrix form

$$(X'W_{r-1}X)\beta_r = X'W_{r-1}z_{r-1}, \tag{5.9}$$

where X is the $(n \times p)$ matrix of explanatory variable values, W_{r-1} is the diagonal matrix whose ith diagonal element is

$$w_{ii} = \left(\frac{\partial \mu_i}{\partial \eta_i}\right)^2 \frac{1}{\text{Var}(Y_i)}$$

evaluated at $\beta = \beta_{r-1}$, and z_{r-1} is the $(n \times 1)$ vector whose ith element is

$$z_i = \sum_{k=1}^{p} x_{ik}\beta_k + (y_i - \mu_i)\left(\frac{\partial \eta_i}{\partial \mu_i}\right)$$

also evaluated at $\beta = \beta_{r-1}$. Equation (5.9) resembles a least squares regression equation (see, e.g., Equation 3.11) for regression of the 'dependent' vector z on the explanatory matrix X, but has an extra matrix W in the middle of each side. The effect of this $(n \times n)$ diagonal matrix is to multiply all elements of the ith column of X' by w_{ii}, for $i = 1, \ldots, n$, on each side of the equation. This multiplication is equivalent to a *weighting* of the ith individual by w_{ii} when doing the regression, so that W can be viewed as a matrix of 'weights'. Furthermore, this weight matrix and the vector z are both updated each time that a new iteration provides an updated value of β. This method of solving the likelihood equations is thus often referred to as *iteratively weighted least squares*.

Having obtained the maximum likelihood estimator $\hat{\beta} = (\hat{\beta}_1, \ldots, \hat{\beta}_p)'$ by one of these iterative schemes, a prime interest is to obtain the *fitted values* $\hat{y}_i$ for $i = 1, \ldots, n$ under the model. The principle is exactly the same as that behind the computation of fitted values in regression (Sections 3.2.3 and 3.5.1), but the computation is slightly more complicated due to the more complex nature of the model. Once again the fitted values are just the expectations under the fitted model, i.e. $\hat{y}_i = \hat{\mu}_i$ for $i = 1, \ldots, n$, where $\hat{\mu}_i$ is the fitted mean, and the fitted linear predictor is again $\hat{\eta}_i = \hat{\beta}_1 x_{i1} + \cdots + \hat{\beta}_p x_{ip}$. Now however the model specifies that $\eta_i = g(\mu_i)$, so we obtain $\hat{y}_i = \hat{\mu}_i = g^{-1}(\hat{\eta}_i)$ for $i = 1, \ldots, n$.

Turning to sampling behaviour of the estimator $\hat{\beta}$, the multivariate versions of standard maximum likelihood properties (see Section 2.5.1) ensure that $\hat{\beta}$

is a consistent estimator of β, and that for large n its sampling distribution is approximately (multivariate) normal with mean β and variance–covariance matrix I^{-1}. In the special case $a_i(\phi) = \phi$ for all i, we can show that

$$I = \frac{1}{\phi} X'WX$$

so that the estimated variance–covariance matrix of $\hat{\beta}$ is $\phi(X'WX)^{-1}$ where W is evaluated at $\hat{\beta}$. However, ϕ is only known for some distributions (e.g. $\phi = 1$ for both the binomial and Poisson, as shown in Examples 5.3 and 5.4). When it is not known it must also be estimated, and a consistent estimator is given by

$$\tilde{\phi} = \frac{1}{n-p} \sum_{i=1}^{n} \frac{(y_i - \hat{\mu}_i)^2}{V(\hat{\mu}_i)}$$

(the divisor $n - p$ allowing for estimation of $\beta_1, \ldots, \beta_p$).

For example, if $Y_1, Y_2, \ldots, Y_n$ are iid $N(\mu, \sigma^2)$ random variables then we see from Example 5.2 that $V(\mu) = 1$ so that (since $\hat{\mu}_i = \hat{y}_i$) we have

$$\tilde{\phi} = \hat{\sigma}^2 = \frac{1}{n-p} \sum_{i=1}^{n} (y_i - \hat{y}_i)^2$$

as expected.

Example 5.6. Suppose that $Y_1, Y_2, \ldots, Y_n$ are Poisson variables, and that we use the canonical link. Then from Example 5.4 we have $a_i(\phi) = 1$ and $b(\theta_i) = e^{\theta_i}$ so that

$$\mu_i = b'(\theta_i) = e^{\theta_i} \quad \text{and} \quad V(\mu_i) = b''(\theta_i) = e^{\theta_i} = \mu_i = \text{Var}(Y_i).$$

The canonical link is given by $g(\mu_i) = \log(\mu_i)$, so that

$$g'(\mu_i) = \frac{1}{\mu_i} = \frac{1}{V(\mu_i)}.$$

Also, if the linear predictor is $\eta_i = \beta_1 x_{i1} + \cdots + \beta_p x_{ip}$, then $\eta_i = g(\mu_i)$ implies that

$$\mu_i = e^{\eta_i} = e^{\beta_1 x_{i1} + \cdots + \beta_p x_{ip}}$$

so that

$$\frac{\partial \mu_i}{\partial \eta_i} = e^{\eta_i} = \mu_i.$$

Direct substitution of these quantities into the expressions given above shows that the elements of z and W in the iterations of the maximum likelihood estimation procedure are given by

$$z_i = \beta_1 x_{i1} + \cdots + \beta_p x_{ip} + \frac{y_i - e^{\beta_1 x_{i1} + \cdots + \beta_p x_{ip}}}{e^{\beta_1 x_{i1} + \cdots + \beta_p x_{ip}}}$$

and

$$w_{ii} = e^{\beta_1 x_{i1} + \cdots + \beta_p x_{ip}},$$

each expression being evaluated at the current values of $\beta_1, \ldots, \beta_p$.

5.4   Assessing the fit of a model: deviance

Having fitted a model to a set of data, the first step of any analysis should always be to inspect the quality of fit provided by the model as a well-fitting model is a necessary prerequisite for reliable inferences. Of course, the more parameters there are in a model, the better will it fit the available data. In the extreme, when the number of parameters equals the number of individuals in the data set, the model is said to be *saturated* and all the fitted values $\hat{y}_i$ match the corresponding observed values y_i exactly. The objective of most statistical modelling is to identify a *small* set of parameters that lead to *good* agreement between observed and predicted values. If the discrepancies between observed and predicted values lie within acceptable limits of (experimental) error, then the model can be said to provide a good fit to the data.

We therefore need to define some suitable statistic that will quantify the overall discrepancy between observed and predicted values. Various possibilities exist, but given that a generalized linear model is defined and fitted using ideas based on the likelihood, it seems sensible to continue with the same concept when assessing the fit of the model also. A natural statistic to use is therefore the generalized likelihood ratio statistic, introduced in Section 2.5.3 for testing hypotheses about parameters in a model.

The idea behind the generalized likelihood ratio test is that we find the maximum of the likelihood of the data twice, once when the null hypothesis about the parameters is assumed to be true and once when no restrictions are imposed on the parameters. If the null hypothesis is indeed true then the two maxima should be very close in value, but if the null hypothesis is not true then the unconstrained maximum should be (much) larger than the constrained maximum. The ratio of the two maxima thus provides a good measure of the 'reasonableness' of the null hypothesis: the closer the ratio is to 1, the more evidence there is in favour of this hypothesis.

In the present case we view the model we are fitting, with p parameters, as if it were the saturated model (that has n parameters) but with $n - p$ of its parameters set to zero. The likelihood ratio statistic for testing the null hypothesis that these (implicit) parameters are zero will thus provide a measure of discrepancy between our model and the saturated model, and hence by implication a measure of goodness of fit of our model to the data.

Expressing the model in canonical form, and writing $\hat{\theta}_i, \tilde{\theta}_i$ for the estimates of the ith canonical parameter under the fitted and saturated models respectively, the generalized likelihood ratio test statistic is given from Section 2.5.3 and Equation (5.4) by

$$\Lambda = \frac{\exp \sum_{i=1}^{n} \{[y_i \hat{\theta}_i - b(\hat{\theta}_i)]/a_i(\phi) + c(y_i, \phi)\}}{\exp \sum_{i=1}^{n} \{[y_i \tilde{\theta}_i - b(\tilde{\theta}_i)]/a_i(\phi) + c(y_i, \phi)\}}$$

$$= \exp \sum_{i=1}^{n} \{[y_i(\hat{\theta}_i - \tilde{\theta}_i) - b(\hat{\theta}_i) + b(\tilde{\theta}_i)]/a_i(\phi)\}.$$

Thus

$$-2\log \Lambda = 2 \sum_{i=1}^{n} [y_i(\tilde{\theta}_i - \hat{\theta}_i) - b(\tilde{\theta}_i) + b(\hat{\theta}_i)]/a_i(\phi), \qquad (5.10)$$

so that in the special case $a_i(\phi) = \phi$ we have $-2\log \Lambda = D/\phi$ where

$$D = 2 \sum_{i=1}^{n} [y_i(\tilde{\theta}_i - \hat{\theta}_i) - b(\tilde{\theta}_i) + b(\hat{\theta}_i)]. \qquad (5.11)$$

D is known as the *deviance* of the model, and D/ϕ is known as the *scaled deviance*. Since this latter quantity is just equal to $-2\log \Lambda$, it provides a measure of discrepancy between the fitted and the saturated models, i.e. between the model and the data. A large scaled deviance indicates a poor fit, while a small scaled deviance indicates a good fit. To decide on what values are 'large' and what values are 'small', we use the asymptotic null distribution of $-2\log \Lambda$ given in Section 2.5.3. If the fitted model is 'correct', then the scaled deviance behaves like a chi-squared variate on $n - p$ degrees of freedom. Thus a calculated scaled deviance that exceeds the upper $100(1 - \alpha)$ percent point of the χ^2_{n-p} distribution indicates a poor fit to the data at the $100\alpha\%$ significance level.

Example 5.7. Continuing Example 5.6, let us suppose that $Y_1, Y_2, \ldots, Y_n$ are independent Poisson variables, and that we use the canonical link. Then we have already established that $a_i(\phi) = 1$, $b(\theta_i) = e^{\theta_i}$, $\mu_i = b'(\theta_i) = e^{\theta_i}$ and that the canonical link is $g(\mu_i) = \log(\mu_i)$. Thus for linear predictor $\eta_i = g(\mu_i) = \beta_1 x_{i1} + \cdots + \beta_p x_{ip}$ we have $\mu_i = e^{\eta_i} = e^{\beta_1 x_{i1} + \cdots + \beta_p x_{ip}}$.

Performing the iterations outlined in Example 5.6 leads to maximum likelihood estimates $\hat{\beta}_1, \ldots, \hat{\beta}_p$ and hence to the fitted means

$$\hat{\mu}_i = e^{\hat{\beta}_1 x_{i1} + \cdots + \hat{\beta}_p x_{ip}}. \qquad (5.12)$$

Also, $\mu_i = e^{\theta_i}$ implies that $\hat{\theta}_i = \log(\hat{\mu}_i)$ and $b(\theta_i) = e^{\theta_i}$ implies that $b(\hat{\theta}_i) = \hat{\mu}_i$.

For the saturated model we have $\tilde{\mu}_i = y_i$, so from the relationships between the parameters given above we obtain $\tilde{\theta}_i = \log(y_i)$ and $b(\tilde{\theta}_i) = y_i$. Thus substituting all these quantities into (5.11) we obtain the scaled deviance of the model as

$$D = 2 \sum_{i=1}^{n} [y_i \{\log(y_i) - \log(\hat{\mu}_i)\} - y_i + \hat{\mu}_i] = 2 \sum_{i=1}^{n} \left[y_i \log\left(\frac{y_i}{\hat{\mu}_i}\right) - (y_i - \hat{\mu}_i) \right]$$

where $\hat{\mu}_i$ is obtained from (5.12).

Example 5.8. Now let us suppose that $Y_1, Y_2, \ldots, Y_n$ are independent binomial variables such that $Y_i \sim B(n_i, \pi_i)$, and we formulate a generalized linear model for predictors $X_1, \ldots, X_p$ using the canonical link.

First, we have density functions

$$f(y_i) = \binom{n_i}{y_i} \pi_i^{y_i} (1 - \pi_i)^{n_i - y_i}$$

and expectations

$$\mu_i = n_i \pi_i$$

for $i = 1, \ldots, n$.

Next, it follows from results in Examples 5.3 and 5.5 that $a_i(\phi) = 1$, $b(\theta_i) = n_i \log(1 + e^{\theta_i})$, $\mu_i = b'(\theta_i) = \frac{n_i e^{\theta_i}}{1 + e^{\theta_i}}$ and the canonical link is

$$g(\mu_i) = \log\left(\frac{\mu_i}{n_i - \mu_i}\right) = \log\left(\frac{\pi_i}{1 - \pi_i}\right).$$

Thus for the linear predictor $\eta_i = g(\mu_i) = \beta_1 x_{i1} + \cdots + \beta_p x_{ip}$ we have

$$\pi_i = \frac{\mu_i}{n_i} = \frac{e^{\eta_i}}{1 + e^{\eta_i}} = \frac{e^{\beta_1 x_{i1} + \cdots + \beta_p x_{ip}}}{1 + e^{\beta_1 x_{i1} + \cdots + \beta_p x_{ip}}}.$$

Performing the iterative maximum likelihood procedure leads us to estimates $\hat{\beta}_1, \ldots, \hat{\beta}_p$, and hence to the fitted means

$$\hat{\mu}_i = n_i \hat{\pi}_i = \frac{n_i e^{\hat{\beta}_1 x_{i1} + \cdots + \hat{\beta}_p x_{ip}}}{1 + e^{\hat{\beta}_1 x_{i1} + \cdots + \hat{\beta}_p x_{ip}}}. \tag{5.13}$$

Also, $\mu_i = \frac{n_i e^{\theta_i}}{1 + e^{\theta_i}}$ and $\mu_i = n_i \pi_i$ together imply that $\hat{\theta}_i = \log(\frac{\hat{\pi}_i}{1 - \hat{\pi}_i})$ and $b(\theta_i) = n_i \log(1 + e^{\theta_i})$ implies that $b(\hat{\theta}_i) = -n_i \log(1 - \hat{\pi}_i)$.

For the saturated model we have $\tilde{\mu}_i = y_i$, i.e. $\tilde{\pi}_i = y_i / n_i$, so from the relationships between the parameters given above we obtain $\tilde{\theta}_i = \log(\frac{y_i}{n_i - y_i})$ and $b(\tilde{\theta}_i) = -n_i \log(\frac{n_i - y_i}{n_i})$. Thus substituting all these quantities into (5.11), and tidying up the algebra, we obtain the scaled deviance of the model as

$$D = 2 \sum_{i=1}^{n} \left[y_i \log\left(\frac{y_i}{\hat{\mu}_i}\right) + (n_i - y_i) \log\left(\frac{n_i - y_i}{n_i - \hat{\mu}_i}\right) \right]$$

where $\hat{\mu}_i$ is obtained from (5.13).

5.5   Comparing models: analysis of deviance

An important question when building statistical models is that of deciding whether all the available explanatory variables are necessary, or whether some of them can be ignored without materially worsening the fit of model to data. In the multiple regression framework the first step in a systematic approach to the problem is to develop a test of the null hypothesis

$$H_0 : \beta_{q+1} = \cdots = \beta_p = 0$$

against the general alternative that at least one of these β_i is not zero, where the full model being fitted is given by

$$y_i = \beta_1 x_{i1} + \beta_2 x_{i2} + \cdots + \beta_q x_{iq} + \beta_{q+1} x_{i,q+1} + \cdots + \beta_p x_{ip} + \epsilon_i$$

with the usual assumptions made about the ϵ_i (and the explanatory variables being permuted if necessary to ensure that the ones under test are the $p - q$ at the end of the sequence). Section 3.4.1 was devoted to this question, and it was shown that the appropriate procedure was to fit the full model (including all explanatory variables), then to fit the reduced model (omitting $X_{q+1}, \ldots, X_p$), and to base the hypothesis test on a comparison of the difference in regression sum of squares between these two models with the residual sum of squares of the full model. The necessary calculations were most conveniently set out in an analysis of variance.

Since the generalized linear model has comparable structure to regression models (but allows greater distributional flexibility), similar considerations of model selection play a part here also. It would also seem entirely reasonable to follow the same general approach to analysis as in the multiple regression case, namely to fit both the 'full' model and the 'reduced' model and then to base a decision on a comparison of the two models. However, since we now allow *any* distribution within the exponential family for the dependent variables, analysis of variance is no longer an appropriate framework for the model comparison (since this technique is only relevant when the variables are normal). Instead, we must turn to the wider framework provided by the generalized likelihood ratio test, and this in turn replaces analysis of variance by analysis of *deviance* for model choice. We now outline the main ideas.

Formally, if we suppose that p explanatory variables $X_1, \ldots, X_p$ are available and we wish to decide whether the last $p - q$ of them, $X_{q+1}, \ldots, X_p$, are necessary when fitting a generalized linear model as specified in Section 5.2.2 above, then we require to test the null hypothesis H_0 given above against the general alternative for the parameters in the linear predictor. The generalized likelihood ratio test statistic for this hypothesis can be written (Section 2.5.3) as

$$\Lambda_0 = \frac{\hat{L}_R}{\hat{L}_F},$$

where $\hat{L}_F$ is the maximized likelihood of the y_i when the full model is fitted (i.e. when all p explanatory variables are included) and $\hat{L}_R$ is the maximized likelihood of the y_i when the reduced model is fitted (i.e. when only the first q explanatory variables are included). If H_0 is true then $W = -2 \log \Lambda_0$ has an approximate χ^2_{p-q} distribution, so large values of W (i.e. those exceeding the chosen critical value of the χ^2_{p-q} distribution) lead to rejection of H_0.

Now if we denote by $\hat{L}_S$ the likelihood of the saturated model then we note that the generalized likelihood ratio statistic above is

$$\Lambda_0 = \frac{\hat{L}_R}{\hat{L}_S} \div \frac{\hat{L}_F}{\hat{L}_S},$$

so that

$$W = -2\left[\log\left(\frac{\hat{L}_R}{\hat{L}_S}\right) - \log\left(\frac{\hat{L}_F}{\hat{L}_S}\right)\right].$$

Thus if we again assume that $a_i(\phi) = \phi$ in the definition of the model, and if we denote the scaled deviances of the full and reduced models by D_F/ϕ and D_R/ϕ respectively, then we see that

$$W = (D_R - D_F)/\phi.$$

For those members of the exponential family for which ϕ is known (e.g. the binomial and Poisson distributions, for both of which we have seen in examples above that $\phi = 1$), the likelihood ratio statistic W is thus obtained directly as the difference of two unscaled deviances and the test of H_0 is conducted by means of the chi-squared test described above.

When ϕ is not known, however, it must be estimated. One possible estimator has already been given, in Section 5.3, but a simpler estimator can be obtained using the method of moments (Section 2.5.1). We know from the previous section that D_F/ϕ has an asymptotic χ^2_{n-p} distribution. This distribution has expected value $n - p$, so by equating observed and expected values we obtain the estimator

$$\tilde{\phi} = D_F/(n - p).$$

Replacing the unknown ϕ by its estimate yields the modified test statistic

$$W = \frac{(D_R - D_F)/(p - q)}{D_F/(n - p)}.$$

If H_0 is true, the numerator has an approximate χ^2_{p-q} distribution, the denominator has a χ^2_{n-p} distribution, and the two distributions are independent. Thus under H_0 W has an $F_{p-q,n-p}$ distribution, which thus forms the basis of the test. Note the equivalent form of this test to that for the corresponding situation in multiple regression, the only difference being that the sums of squares of the latter are replaced by the deviances of the present test.

We can now see an immediate extension of this test for comparison of a number of nested models. Suppose that we have a sequence of models, specified as sets of linear predictor parameters as

$$M_1 \supset M_2 \supset M_3 \supset \cdots \supset M_k.$$

For example, M_1 might be the full model including all β_i, and then $M_2, M_3, \ldots$ are obtained by successively omitting one or more of the β_i present in the 'current' model. If we calculate the deviances $D_1, D_2, D_3, \ldots$ of each model, and take successive differences $D_2 - D_1, D_3 - D_2, \ldots$, then we can use the above theory to assess which of these differences are significant and which are not, thus building up a picture of 'good' models to fit to the data. This generalizes the analysis of variance of previous chapters to an *analysis of deviance* for use with generalized linear models. Of course, it must be stressed that the exact normal-distribution theory of analysis of variance is now replaced by *approximate* chi-squared distribution theory for differences of deviances. Moreover, in general there will be non-orthogonality of terms in an analysis of deviance table, so care must be exercised in any interpretations or conclusions reached. Generally speaking, several different nested model sequences, each with its own analysis of deviance table, must be considered before any firm decisions are reached. However, such complications notwithstanding, the framework is now available for modelling a huge variety of situations.

Example 5.9. Table 5.1 shows the numbers of subjects (Y) out of samples of size N that tested positive for toxoplasmosis in 34 cities in El Salvador, together with the amount of rainfall (X) in mm measured in each city. If interest lies in investigating whether there is any relationship between the amount of rainfall in a city and the proportion of subjects that tested positive for toxoplasmosis, then this is an appropriate situation for fitting a generalized linear model. The observed number y_i testing positive in city i may obviously be assumed to have a binomial distribution with parameters n_i, the sample size in that city, and π_i, the probability that any individual from that city tests positive. The latter parameter will differ from city to city and may be linked to the rainfall value x_i in a city via the linear predictor. To allow a range of possibilities, let us specify a polynomial of order p in the rainfall value as the linear predictor:

$$\eta_i = \beta_0 + \beta_1(x_i - \overline{x}) + \beta_2(x_i - \overline{x})^2 + \cdots + \beta_p(x_i - \overline{x})^p$$

where $\overline{x}$ is the mean rainfall over the cities. Finally, although there are various functions that could be chosen for the link function, the logit link has been shown in Example 5.5 to be the canonical link and so is the most convenient choice. We thus investigate the model

$$\log\{\pi_i/(1 - \pi_i)\} = \beta_0 x_{i0} + \beta_1(x_i - \overline{x}) + \beta_2(x_i - \overline{x})^2 + \cdots + \beta_p(x_i - \overline{x})^p,$$

where the observations are from a binomial distribution.

Fitting the model for any chosen order p of polynomial, and obtaining its deviance (which, since $\phi = 1$ for the binomial, is also the scaled deviance) proceeds as described in the sections above. Deviances for all orders from 0 (a constant) up to 4 (a quartic) are as follows, together with the successive decreases in deviances each time the order is increased by 1:

Table 5.1 *Number of subjects sampled and testing positive for toxoplasmosis in 34 cities in El Salvador, with rainfall in mm for each city (Source: Efron, 1986).*

City (i)	Number tested (N)	Number positive (Y)	Rainfall (X)
1	4	2	1735
2	10	3	1936
3	5	1	2000
4	10	3	1973
5	2	2	1750
6	5	3	1800
7	8	2	1750
8	19	7	2077
9	6	3	1920
10	10	8	1800
11	24	7	2050
12	1	0	1830
13	30	15	1650
14	22	4	2200
15	1	0	2000
16	11	6	1770
17	1	0	1920
18	54	33	1770
19	9	4	2240
20	18	5	1620
21	12	2	1756
22	1	0	1650
23	11	8	2250
24	77	41	1796
25	51	24	1890
26	16	7	1871
27	82	46	2063
28	13	9	2100
29	43	23	1918
30	75	53	1834
31	13	8	1780
32	10	3	1900
33	6	1	1976
34	37	23	2292

Order of polynomial	Deviance	Degrees of freedom	Decrease in deviance
0	74.19	33	
1	74.06	32	0.13
2	74.06	31	0.00
3	62.61	30	11.45
4	62.42	29	0.19

We see very little difference in deviance values until the introduction of the cubic term, when the deviance drops by 11.45 (significant at the 0.1% level against the χ_1^2 distribution). The deviance value once again stabilizes beyond third order, and no higher-order polynomial gives a significant drop in deviance. Comparing the cubic polynomial with the constant model (i.e. the hypothesis of no relationship between rainfall and probability of testing positive) gives a difference in deviance of $74.19 - 62.61 = 11.58$, which is significant at the 1% level against the χ_3^2 distribution, so this cubic seems to be the best model of polynomial form for these data. For rainfall values in metres (i.e. tabulated $x_i \div 1000$), the parameter estimates and their standard errors under the binomial assumption are:

Parameter	Estimate	Std. Error
β_0	0.0996	0.102
β_1	−2.55	0.883
β_2	−6.06	2.96
β_3	39.3	11.7

and a plot of the fitted response superimposed on a scatter diagram of the data is shown in Fig. 5.1. For further discussion of this example, and details of the analysis, see Chapter 3 by D. Firth in *Statistical Theory and Modelling* (1991, Chapman & Hall, edited by D.V. Hinkley, N. Reid and E.J. Snell).

5.6 Normal models

We have stated in Section 5.4 that the scaled deviance behaves asymptotically like a chi-squared variate if the fitted model is 'correct', and in Section 5.5 that the difference in scaled deviances for two nested models also has an asymptotic chi-squared distribution under the null hypothesis that the extra parameters in the 'larger' of the two models are all zero. It is thus common practice to *approximate* the distributions of these two quantities by the appropriate chi-squared forms. Two comments need to be made about this practice. First, theoretical considerations show that the chi-squared approximation of the scaled deviance itself can sometimes be very poor, but the difference of scaled deviances for nested models is much more accurately approximated by a chi-squared distribution under the null hypothesis. Second, in the special case of normal models,

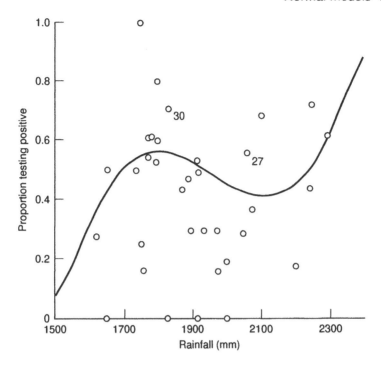

Fig. 5.1 *Toxoplasmosis data: proportion testing positive against rainfall, plus superimposed fitted cubic (from Hinkley et al 1991, Chapter 3).*

these chi-squared distributions are actually the *exact* distributions of the two quantities. Thus hypothesis tests in generalized linear models are best restricted to those involved in differences of scaled deviances. In general such tests are only approximate, but in the special case of normality of response variables they are exact.

We further commented in Section 5.5 that the analysis of deviance for a sequence of nested models was very reminiscent of the analysis of variance breakdown for a sequence of nested multiple regression models. We now show that it is not merely very reminiscent of this analysis (described in Chapter 3), but indeed is exactly the same analysis if the response variable is assumed to be normally distributed.

First, recollect from (3.10) that the matrix form of the multiple regression model is given by

$$y = X\beta + \epsilon \tag{5.14}$$

where ϵ has mean vector $\mathbf{0}$ and covariance matrix $\sigma^2 I$. For consistency with Chapter 3 we separate the constant term from the p explanatory variables in this model, thereby producing $p + 1$ elements of β (whereas in Sections 5.4 and 5.5

we assumed there were p elements). Now if we add the assumption of normality about ϵ then we can recognize this model as a generalized linear model with normal errors, linear predictor given by the multiple regression equation, and identity link (as shown at the start of Section 5.2). Following Section 5.2.1 on the exponential family, and Example 5.2 on the normal distribution, we see that σ^2 is the scale parameter and that emphasis focusses on estimation of β. Thus we have the likelihood

$$L(\beta, \sigma^2) = \frac{1}{(2\pi\sigma^2)^{p/2}} \exp\left\{-\frac{1}{2\sigma^2}(y - X\beta)'(y - X\beta)\right\}, \qquad (5.15)$$

and it is evident that to maximize this likelihood with respect to β we need to find the value of β that minimizes

$$S(\beta) = (y - X\beta)'(y - X\beta).$$

This is, of course, precisely the same β value that provides the least squares estimator, i.e.

$$\hat{\beta} = (X'X)^{-1}X'y$$

from (3.12), so that in this case the maximum likelihood estimation does not need an iterative scheme. The maximized value of the likelihood is thus

$$L(\hat{\beta}, \sigma^2) = \frac{1}{(2\pi\sigma^2)^{p/2}} \exp\left\{-\frac{1}{2\sigma^2}(y - X\hat{\beta})'(y - X\hat{\beta})\right\}.$$

Next, for the saturated model we have as usual $X\tilde{\beta} = y$ and the maximized likelihood

$$L(\tilde{\beta}, \sigma^2) = \frac{1}{(2\pi\sigma^2)^{p/2}}.$$

Hence we obtain the scaled deviance

$$D_1 = \frac{1}{\sigma^2}(y - X\hat{\beta})'(y - X\hat{\beta}),$$

which reduces to

$$D_1 = \frac{1}{\sigma^2}(y'y - \hat{\beta}X'y)$$

on substituting for $\hat{\beta}$ and simplifying the algebra. We note that this deviance is just $\frac{1}{\sigma^2}SS_E$, where SS_E is the residual sum of squares in the multiple regression ANOVA, and hence we see from Section 3.3.4 that D has (exactly) a chi-squared distribution on $n - p - 1$ degrees of freedom.

The same argument establishes that the deviance D_2 of a reduced model

$$y = X_1\beta_1 + \epsilon$$

(which has a constant term plus q of the p original explanatory variables) has a chi-squared distribution on $n - q - 1$ degrees of freedom and hence that the difference of deviances $D_2 - D_1$ has a chi-squared distribution on $p - q$ degrees of freedom. Moreover, from the form above it is evident that this difference of deviances equals

$$D_2 - D_1 = \frac{1}{\sigma^2}(\hat{\beta}X'y - \hat{\beta}_1 X_1'y),$$

which is $\frac{1}{\sigma^2}$ times the extra sum of squares due to the omitted regressors (Section 3.4.1). It thus follows that the significance of these regressors can be tested by means of the ratio

$$W = \frac{(D_2 - D_1)/(p - q)}{D_1/(n - p - 1)},$$

which has an $F_{p-q,n-p-1}$ distribution on the null hypothesis. This is exactly the same test as was obtained in the previous analysis of variance formulation, so the equivalence of the two forms of analysis has been shown.

5.7 Inspecting and checking models

As has already been stressed several times in this book, any model is only an approximation to reality, all models involve various assumptions about the data to which they relate, and some portions of the data may be more consonant with the model than others. It is thus essential to check all aspects of a model critically, especially when the fitted model is to be used later for analysis or prediction.

The topic of model validation and criticism has been studied extensively in the case of multiple regression, and we summarized in Section 3.5 some of the more common statistics that are used to inspect individual observations for possible deviation from model assumptions. In particular, we focussed on *residuals* as a means of screening potential outliers in the data set, and gave several measures of *influence* to identify observations whose removal from the data would cause large changes in the fitted model.

Not unnaturally, given the extensive experience with such quantities when fitting models to normal variables, statisticians have sought to use analogous measures in the case of generalized linear models also. Broadly speaking their efforts have been successful and the general techniques are essentially those outlined in Section 3.5. However, for generalized linear models we need to reconsider slightly the definition of a residual, in order to cope with all the distributions that may replace the normal. We thus summarize below the main changes; the resulting quantities may then be used exactly as if they were the standard normal residuals of Chapter 3. As has already been mentioned in Section 5.3 above, we take the fitted value $\hat{y}_i$ to be the estimated mean $\hat{\mu}_i$ for a generalized linear model.

The ordinary residual for the ith observation y_i is just the difference between the observed value y_i and the fitted value $\hat{\mu}_i$. To allow for differential variability, we should scale this quantity by the standard error of y_i. Generally, this standard error will be a function of model parameters so needs to be estimated, and the simplest procedure is to replace these parameters in the expression by their estimates. We will denote this estimated standard error by $\sqrt{\widehat{\text{var}(y_i)}}$. For the same reasons as in the normal case, we should additionally make allowance for the effect of the explanatory variables on the residual and divide also by $\sqrt{(1 - h_i)}$, where h_i is the ith diagonal element of

$$H = W^{1/2}X(X'WX)^{-1}X'W^{1/2}$$

and W is the value of the weight matrix defined in (5.9) at the conclusion of the iterations for maximum likelihood estimation in the model. This matrix H is of course the generalized linear model equivalent of the multiple regression 'hat' matrix defined in Section 3.5, arising from the matrix expression (5.9) of the generalized linear model estimating equations. For historical reasons the residuals in generalized linear models are referred to as *Pearson* residuals so, finally, the *standardized Pearson residual* is defined to be

$$r_{i(P)} = \frac{y_i - \hat{\mu}_i}{\sqrt{[(1 - h_i)\widehat{\text{var}(y_i)}]}}. \tag{5.16}$$

Various other residuals have also been proposed from time to time. The two most useful ones are the *standardized deviance residual*

$$r_{i(D)} = \frac{\text{sgn}(y_i - \hat{\mu}_i)\sqrt{D_i}}{\sqrt{(1 - h_i)}} \tag{5.17}$$

where D_i is the contribution made to the scaled deviance by observation i (with ϕ estimated if necessary) and $\text{sgn}(y_i - \hat{\mu}_i)$ is the sign of $(y_i - \hat{\mu}_i)$; and the *likelihood residual*

$$r_{i(L)} = \text{sgn}(y_i - \hat{\mu}_i)\sqrt{h_i r_{i(P)}^2 + (1 - h_i)r_{i(D)}^2}. \tag{5.18}$$

The standardized deviance residuals pin-point any observations that give a disproportionately large contribution to this deviance, while the usefulness of the likelihood residuals comes from the fact that $r_{i(L)}^2$ is approximately equal to the change in scaled deviance that is caused by removal of the ith observation from the data. The likelihood residuals thus provide measures of influence of the observations, and a modified Cook's distance (Section 3.5) can be defined by

$$C_i = |r_{i(L)}|\sqrt{\frac{(n - p)h_i}{p(1 - h_i)}}. \tag{5.19}$$

In addition to their use as diagnostics to highlight problematic observations, the standardized Pearson residuals $r_{i(P)}$ provide an alternative to the scaled deviance for measuring the overall lack of fit of the model. Their sum of squares $X^2 = \sum_{i=1}^{n} r_{i(P)}^2$ produces the well-known Pearson X^2-statistic, which has the same asymptotic chi-squared distribution as the scaled deviance if the fitted model is 'correct'.

Example 5.10. Returning to the toxoplasmosis data of Example 5.9, let us compute and compare the various quantities defined above.

Given that a cubic polynomial in rainfall provided the most appropriate fit, we have the following model elements:

$$\mu_i = n_i \pi_i = \frac{n_i e^{\eta_i}}{1 + e^{\eta_i}},$$

for the linear predictor

$$\eta_i = \beta_0 + \beta_1(x_i - \bar{x}) + \beta_2(x_i - \bar{x})^2 + \beta_3(x_i - \bar{x})^3$$

$(i = 1, \ldots, 34)$ and, since errors are binomial, $\text{Var}(Y_i) = n_i \pi_i (1 - \pi_i)$.
We thus have

$$\frac{\partial \mu_i}{\partial \eta_i} = \frac{n_i e^{\eta_i}}{1 + e^{\eta_i}} - \frac{n_i e^{2\eta_i}}{(1 + e^{\eta_i})^2}$$

$$= \mu_i - \frac{\mu_i^2}{n_i}$$

$$= \mu_i \left(1 - \frac{\mu_i}{n_i}\right)$$

$$= n_i \pi_i (1 - \pi_i),$$

so that from (5.9) the diagonal elements of the matrix W are given by

$$w_{ii} = [n_i \pi_i (1 - \pi_i)]^2 \frac{1}{n_i \pi_i (1 - \pi_i)} = n_i \pi_i (1 - \pi_i).$$

Using the parameter estimates given in Example 5.9, and taking the rainfall values x_i in metres, we have the fitted probabilities

$$\hat{\pi}_i = \frac{e^{[0.0996 - 2.55(x_i - \bar{x}) - 6.06(x_i - \bar{x})^2 + 39.3(x_i - \bar{x})^3]}}{1 + e^{[0.0996 - 2.55(x_i - \bar{x}) - 6.06(x_i - \bar{x})^2 + 39.3(x_i - \bar{x})^3]}},$$

and hence can obtain the fitted values $\hat{\mu}_i$ and the estimated elements $\hat{w}_{ii}$ by direct substitution in the formulae above. The latter quantities enable us to find the diagonal elements h_i of the hat matrix (since we know X), so that we have all the quantities needed for calculation of the Pearson residuals $r_{i(P)}$ from (5.16). Moreover, from the deviance given at the end of Example 5.8 we see that

$$D_i = 2\left[y_i \log\left(\frac{y_i}{\hat{\mu}_i}\right) + (n_i - y_i) \log\left(\frac{n_i - y_i}{n_i - \hat{\mu}_i}\right)\right],$$

Table 5.2 *Pearson residuals* $r_{i(P)}$, *deviance residuals* $r_{i(D)}$, *likelihood residuals* $r_{i(L)}$ *(all standardized) and modified Cook's distance* C_i *for each city* (i) *in the toxoplasmosis data of Table 5.1.*

i	$r_{i(P)}$	$r_{i(D)}$	$r_{i(L)}$	C_i
1	−0.158	−0.158	−0.158	0.064
2	−1.321	−1.339	−1.338	0.724
3	−1.182	−1.232	−1.231	0.551
4	−1.156	−1.175	−1.175	0.692
5	1.286	1.554	1.553	0.419
6	0.166	0.167	0.167	0.065
7	−1.724	−1.744	−1.743	0.950
8	−0.499	−0.503	−0.502	0.555
9	−0.084	−0.084	−0.084	0.034
10	1.531	1.601	1.599	0.893
11	−1.482	−1.513	−1.510	1.790
12	−1.131	−1.284	−1.284	0.218
13	1.183	1.175	1.179	3.411
14	−2.788	−2.926	−2.905	3.949
15	−0.927	−1.114	−1.113	0.221
16	−0.085	−0.085	−0.085	0.052
17	−1.036	−1.208	−1.208	0.200
18	0.863	0.868	0.867	1.266
19	−0.386	−0.386	−0.386	0.376
20	−0.905	−0.923	−0.915	2.691
21	−2.742	−2.820	−2.817	1.860
22	−0.859	−1.053	−1.050	0.416
23	1.443	1.475	1.471	1.715
24	−0.618	−0.616	−0.617	1.075
25	−1.000	−0.998	−0.999	1.248
26	−0.888	−0.886	−0.886	0.594
27	3.263	3.240	3.250	9.125
28	2.096	2.085	2.086	1.938
29	0.231	0.232	0.232	0.266
30	2.883	2.942	2.930	4.808
31	0.405	0.407	0.407	0.267
32	−1.475	−1.488	−1.488	0.778
33	−1.540	−1.617	−1.616	0.739
34	0.250	0.250	0.250	1.367

which enables us to obtain the deviance residuals $r_{i(D)}$ and the likelihood residuals $r_{i(L)}$ from (5.17) and (5.18) respectively. Finally, we have $n = 34$ and $p = 4$ (constant term plus three powers of x_i) for the fitted model, which yields the modified Cook statistic C_i from (5.19).

These quantities were all calculated for the 34 cities, and the results are shown in Table 5.2. It can be seen that all these statistics provide a very consistent picture. The city whose presence/absence has highest effect on the fitted model, as measured by Cook's distance, is city number 27, and this is the city with highest residuals (of all types); note that this city has a large value of n_i, 82 individuals having been tested there. The two next most influential cities are numbers 30 (with $n_i = 75$) and 14 (with $n_i = 22$) respectively, and these are also the cities with the two next highest sets of residuals. City number 21 has moderately large residuals but a small value of C_i, and none of the other cities warrants particular attention.

Summing the squares of the standardized Pearson residuals yields $X^2 = 68.75$, which can be compared with the scaled deviance value of 62.61 given in Example 5.9. Under the null hypothesis that the model is correct, both quantities have an asymptotic chi-squared distribution on 30 degrees of freedom, so both statistics agree that the fit is not a good one. The problem is essentially one of overdispersion (to be discussed in Chapter 8). Removal of the four individuals identified above and refitting the model will provide a short-term solution, in that the fit is then much improved and the interpretation of the results is facilitated. Allowing for the overdispersion in the fitted model is more satisfactory in general, however.

5.8 Software

Regression and analysis of variance are such well-established and popular techniques that appropriate software for conducting analyses is very readily obtainable. For example, virtually every computing package that offers statistical facilities will include procedures or commands for conducting regression or analysis of variance in one form or another.

By contrast, generalized linear models are somewhat more specialized so appropriate software may be harder to find. The tailor-made package for generalized linear models is GLIM, and this package will handle not only all the models discussed in this book but also many others. A good text for the beginner is *GLIM: An Introduction* by M.J.R. Healy (1989, Oxford University Press), while *Statistical Modelling in GLIM* by M. Aitkin, D. Anderson, B. Francis and J. Hinde (1989, Oxford University Press) tailors a comprehensive discussion of modelling techniques to this package. The facilities available in GLIM are also all available in GENSTAT, along with a very wide selection of other statistical procedures, so this latter package is probably the best one from a general modelling perspective. Other general statistical packages are less well endowed as regards generalized linear models. SAS has a reasonable selection of procedures, but BMDP and SPSS each have rather fewer options so provide more limited possibilities.

S-plus is now fast becomming a powerful and flexible option, mixing built-in modelling procedures with more general programming facilities. The texts by B. Everitt (*A Handbook of Statistical Analyses using S-plus*, 1994, Chapman & Hall) and W.N. Venables and B.D. Ripley (*Modern Applied Statistics with S-plus*, 1995, Springer) provide excellent introductions for the practitioner. Of course, computer software is constantly being updated and improved, so reference should be made to the relevant manuals to check on the exact facilities offered by any specific release of a given package.

6
Binomial response variables: logistic regression and related methods

6.1 Binary response data

One of the simplest, and arguably most common, types of observation in practice is when each each individual can be classified into one of just two possible categories. For example, when inspecting the quality of items in a manufacturing process each item can be categorized as either *satisfactory* or *unsatisfactory*; insects in a toxicology experiment can be classified as *alive* or *dead* a given time after exposure to a toxin; and deliveries in the labour ward of a hospital are deemed to be either *full-term* or *premature*. Sometimes, such a dichotomy is created from an underlying quantitative measurement for reasons of convenience, so a bank might classify a customer as either *good risk* or *bad risk* depending on the balance in that customer's account, and a subject in a psychological experiment might be designated either *normal* or *slow* depending on his or her speed of reaction to an applied stimulus.

To handle such a variable, it is convenient to label one of the categories (usually the 'positive' one towards which the analysis is angled) as *success* and the other as *failure*, and to score these categories 1 and 0 respectively. A typical data set will thus have a series of ones and zeros as the response variable Y. Associated with each individual's Y value there will often be observations on a set of explanatory variables $X_1, X_2, \ldots, X_p$. For example, the bank will have information on each customer's earnings and monthly commitments; the researchers conducting the psychological experiment will typically have information such as age, gender, occupation and so on for each subject; and each mother in the hospital labour ward will have had a whole battery of measurements such as blood pressure, temperature, pulse rate etc regularly taken on her, as well as providing demographic and personal information on herself. In contrast to normal-based regression methods in which we wish to predict the *value* of Y given values for the explanatory variables, interest now centres on predicting the *probability* π that $Y = 1$ given values for the explanatory variables.

It often happens that a *group* of individuals are either treated in the same manner or come from the same homogeneous background, so that they all have the same values for the explanatory variables. For example, in the toxicology experiment, groups of insects may be subjected to varying doses of toxin in such a way that there is a group of insects exposed to each dose. Similarly, the subjects in the psychological experiment may be grouped by age and gender to give several subjects in each grouping. Data of this form are known as *grouped binary data*, and for conciseness it is sufficient simply to record the number of successes and the total number of observations in each group rather than list the individual values of Y. If we have scored Y as 1 and 0 for success and failure respectively, then the total number of successes is given by $\sum Y$. Moreover, if each individual observation is treated as the outcome of a Bernoulli trial and the probability of success remains constant within a group, the Bernoulli distribution is appropriate for individual Y values and the binomial distribution is appropriate for values of $\sum Y$. These distributions were introduced in Section 2.3.1, from which it can be seen that likelihoods obtained from the two distributions will be the same apart from a constant of proportionality (equal to the binomial combinatorial term). Moreover, the Bernoulli distribution is a special case of the binomial distribution. Thus methods for modelling and analysing binary responses are usually based on the assumption of a binomial distribution, and it is with such methods that we are concerned in this chapter.

In Chapter 5 we considered the fundamental aspects of generalized linear models, and gave some specific results for binary response variables as examples of the general theory. Here we take up more detailed aspects of such response variables, and consider a number of practical issues connected with their modelling and analysis.

6.2 Modelling binary response probabilities

For generality, let us focus on the grouped binary case and suppose that we have g groups with n_i individuals in the ith group; each individual in the ith group has the value $x_i = (x_{i1}, x_{i2}, \ldots, x_{ip})'$ on the vector $X = (X_1, X_2, \ldots, X_p)'$ of explanatory variables; and r_i of the individuals exhibit the response 'success', i.e. have value $Y = 1$, while the remaining $n_i - r_i$ exhibit the response 'failure', i.e. have the value $Y = 0$ for $i = 1, \ldots, g$. Let the number of individuals in the sample be $n (= n_1 + \cdots + n_g)$. Ungrouped binary data can be handled within this framework simply by setting $n_i = 1$ for all i, in which case $g = n$. Interest usually focusses on modelling π_i, the probability that $Y = 1$ in the ith group, as a function of x_i.

We have already mentioned in Chapter 5 the main drawback of modelling this probability as a linear function of the explanatory variables, and hence of using the regression methods of Chapter 3 for analysis, namely that any probability is restricted to take values between 0 and 1 but a linear model can give rise to any value between $-\infty$ and $+\infty$. It is thus necessary to transform π_i into a quantity that *does* take values in the interval $(-\infty, +\infty)$ before a linear model can be sensibly applied. The main contenders for such a transformation are:

(i) **The logistic transformation**

$$\xi_i = \log\{\pi_i/(1 - \pi_i)\},$$

often denoted by $\xi_i = \text{logit}(\pi_i)$.

(ii) **The probit transformation**

$$\xi_i = \Phi^{-1}(\pi_i),$$

where $\Phi(\cdot)$ is the cumulative normal distribution function (see Section 2.3.2). Thus ξ_i is the solution of

$$\int_{-\infty}^{\xi_i} \phi(u)\,du = \pi_i$$

where $\phi(u)$ is the standard normal density function $\frac{1}{\sqrt{2\pi}}\exp(-\frac{1}{2}u^2)$.

(iii) **The complementary log–log transformation**

$$\xi_i = \log[-\log(1 - \pi_i)].$$

All these transformations produce sigmoidal curves mapping a probability from an interval $(0, 1)$ to the real line $(-\infty, +\infty)$. The logistic and the probit functions are both symmetric about the value $\pi_i = 0.5$ and are quite similar to each other, but the complementary log–log function is asymmetric and rather different from the others (although it is barely distinguishable from the logistic at small values of π_i). In principle any of these functions can be used as the basis for a link function in a generalized linear model. Ease of computation, interpretation as the logarithm of the odds in favour of a success, and the fact that it provides a canonical link, all favour the use of the logistic transformation in general practical applications. Historically, however, the probit transformation has been used most extensively in bioassay while the complementary log–log transformation finds a use whenever underlying concepts of survival or extremeness are implicitly present. We consider these particular cases briefly towards the end of this chapter, but the main focus is on general-purpose methods based on the logistic transformation.

6.3 Logistic regression

6.3.1 The model

We follow on with the grouped binary situation as specified at the beginning of the previous section, so that the data consist of g groups (i.e. observations) in each of which the explanatory variables are constant across individuals. Changing notation slightly, let the response variable Y_i be the number of 'successes' observed in group i (thus replacing the previous sum $\sum Y$ of the 0/1 responses over the individuals in this group), let the count variable N_i give the total number of individuals in group i, and let the vector $x_i = (x_{i1}, x_{i2}, \ldots, x_{ip})'$ specify the common values of the explanatory variables $X_1, X_2, \ldots, X_p$ for all individuals in group i. As before, π_i is the probability that a success is observed on any individual in group i. We choose the logistic transformation described in the previous section for stretching the scale of these probabilities from $(0, 1)$ to $(-\infty, +\infty)$. In terms of the general specification of Section 5.2.2, therefore, we have a generalized linear model with the following components

(i) The Y_i $(i = 1, \ldots, g)$ are independent binomial random variables with parameters N_i, π_i (so that $\mu_i = N_i \pi_i$).

(ii) The explanatory variables provide a set of linear predictors $\eta_i = \beta_1 x_{i1} + \beta_2 x_{i2} + \cdots + \beta_p x_{ip}$ for $i = 1, \ldots, g$.

(iii) The link between (i) and (ii) is given by $g(\pi_i) = \eta_i$, where $g(\pi_i) = \log\left(\frac{\pi_i}{1-\pi_i}\right)$.

6.3.2 Fitting the model and goodness of fit

From the above model specification, $\log\left(\frac{\pi_i}{1-\pi_i}\right) = \eta_i = \beta_1 x_{i1} + \cdots + \beta_p x_{ip}$ so it follows that

$$\pi_i = \frac{\mu_i}{N_i} = \frac{e^{\eta_i}}{1 + e^{\eta_i}} = \frac{e^{\beta_1 x_{i1} + \cdots + \beta_p x_{ip}}}{1 + e^{\beta_1 x_{i1} + \cdots + \beta_p x_{ip}}}.$$

Thus, taking the two central terms of this expression and differentiating with respect to η_i, we obtain

$$\frac{1}{N_i} \frac{\partial \mu_i}{\partial \eta_i} = \frac{e^{\eta_i}}{(1 + e^{\eta_i})^2}.$$

But since $\pi_i = \frac{e^{\eta_i}}{1+e^{\eta_i}}$ then $1 - \pi_i = \frac{1}{1+e^{\eta_i}}$ and so $\pi_i(1 - \pi_i) = \frac{e^{\eta_i}}{(1+e^{\eta_i})^2}$. Hence

$$\frac{\partial \mu_i}{\partial \eta_i} = N_i \pi_i (1 - \pi_i).$$

Moreover, since the distribution of Y_i is binomial (N_i, π_i), then $\mathrm{Var}(Y_i) = N_i \pi_i (1 - \pi_i)$ also.

Maximum likelihood estimation of the parameters $\beta_1, \beta_2, \ldots, \beta_p$ is effected by means of the iterative scheme outlined in Section 5.3. This iterative scheme requires evaluation of elements of the matrix W and the vector z specified in equation (5.9) at 'current' estimates $\tilde{\beta}_i$ of the parameters $(i = 1, \ldots, p)$. This matrix and vector involve the quantities $\mathrm{Var}(Y_i)$, μ_i and $\frac{\partial \mu_i}{\partial \eta_i}$ in addition to x_{ij}, Y_i and β_i. However, we see from all the expressions above that, given 'current' values $\tilde{\beta}_i$ for all the parameters, then successive substitution yields 'current' values $\tilde{\mu}_i, \tilde{\pi}_i$ and hence those of $\mathrm{Var}(Y_i)$ and $\frac{\partial \mu_i}{\partial \eta_i}$. Thus all necessary quantities are available for the iterative scheme to proceed.

Performing this iterative maximum likelihood procedure leads us to estimates $\hat{\beta}_1, \ldots, \hat{\beta}_p$, and hence to the fitted means

$$\hat{\mu}_i = N_i \hat{\pi}_i = \frac{N_i e^{\hat{\beta}_1 x_{i1} + \cdots + \hat{\beta}_p x_{ip}}}{1 + e^{\hat{\beta}_1 x_{i1} + \cdots + \hat{\beta}_p x_{ip}}}. \tag{6.1}$$

Recollecting from Example 5.3 that the scale parameter for the binomial distribution is 1, we see from Section 5.3 that the approximate variance–covariance matrix for the estimates $\hat{\beta}_1, \ldots, \hat{\beta}_p$ is given by $(X'WX)^{-1}$. Standard errors of the estimates are therefore given by the square roots of the diagonal elements of this matrix. Furthermore, it follows from Example 5.8 that the scaled deviance of the above model is

$$D = 2 \sum_{i=1}^{g} \left[Y_i \log\left(\frac{Y_i}{\hat{\mu}_i}\right) + (N_i - Y_i) \log\left(\frac{N_i - Y_i}{N_i - \hat{\mu}_i}\right) \right]$$

where $\hat{\mu}_i$ is obtained from (6.1). D has an asymptotic chi-squared distribution on $n - p$ degrees of freedom under the null hypothesis that the data could have been generated by this model, but it has already been stressed that this distribution is often a poor approximation to the true null distribution so should be used with caution. An alternative measure of lack of fit is Pearson's X^2-statistic introduced in Section 5.7, but a better procedure is to investigate different (nested) models via analysis of deviance in order to arrive at a 'good' model.

6.3.3 Comparison of models

The general theory has been set out in Section 5.5, and there is little to add here. Suppose that M_1 denotes a model with p parameters $\beta_1, \ldots, \beta_p$ while M_2 is the same model but with q of these parameters set to zero. Then M_2 is nested within M_1 (i.e. $M_1 \supset M_2$), so that if we fit both models as detailed above and obtain their respective deviances D_1, D_2, the difference $D_2 - D_1$ has an approximate chi-squared distribution on q degrees of freedom under the null hypothesis that the omitted parameters really do equal zero. The same result is true more generally for *any* pair of nested models, in which the 'larger' model has q parameters more than the 'smaller' model and the null hypothesis specifies equivalence of the two models. Hence we can test this null hypothesis by referring the calculated value for the difference of deviances to the χ_q^2 distribution: if it exceeds the chosen critical value of this distribution then we reject the null hypothesis and conclude that the larger model must be retained, but if it does not exceed the critical value then we can conclude that the simpler model is as good as the more complicated one. By specifying a sequence of appropriate nested models and conducting such tests on them, we can arrive at a good (i.e. parsimonious) model to fit to the data in analogous fashion to the choice of model in regression analysis (Section 3.4).

Example 6.1. Data relating to factors affecting the risk of heart problems were collected by the Health Promotion Research Trust in 1986. The data consist of frequencies of males reporting some form of 'heart trouble', together with the numbers at risk and their classification with respect to five explanatory factors (social class, smoking, alcohol, family history of heart trouble, and age). The factor levels recorded were:

social class: 'SC', coded 1 for non-manual and 2 for manual;
smoking: 'Smoke', coded 1 for non-smoker, 2 for ex-smoker and 3 for current smoker;
alcohol: 'Alc', coded 1 for non-drinker and 2 for someone who drinks;
family history: 'Hist', coded 1 for no family history and 2 if there is a family
  history of heart trouble;
age: 'Age', coded 1 if age < 40, 2 if age is between 40 and 59, and 3 if age is 60+.

The data are shown in Table 6.1, where N indicates the number of males 'at risk' in each category and Y is the corresponding number reporting heart trouble. We wish to find a model that will explain the proportions at risk in each category in terms of the corresponding levels of the five factors.

Since these factors are all qualitative, we need to introduce dummy binary variables in order to incorporate them in a predictive equation (see, e.g., Section 3.3.8 for corresponding treatment of factors in regression). The full set of binary variables necessary, together with their codings, is as follows.

overall mean: X_1, which takes value 1 for all observations;
SC: X_2, with values 0 for non-manual and 1 for manual;
Smoke: X_3 and X_4, with values $X_3 = X_4 = 0$ for non-smoker, $X_3 = 1$, $X_4 = 0$
  for ex-smoker and $X_3 = 0$, $X_4 = 1$ for current smoker;
Alc: X_5, with values 0 for non-drinker and 1 for drinker;
Hist: X_6, coded 0 for no family history and 1 for family history;
Age: X_7 and X_8, with values $X_7 = X_8 = 0$ if age < 40, $X_7 = 1$, $X_8 = 0$ if
  age is 40–59 and $X_7 = 0$, $X_8 = 1$ if age is 60+.

The full model then has linear predictor $\eta = \beta_1 X_1 + \beta_2 X_2 + \cdots + \beta_8 X_8$ so that, for example, $\eta = \beta_1$ when each factor is at its first level, $\eta = \beta_1 + \beta_4$ when each factor except 'Smoke' is at its first level but the group under consideration is that of current smokers, and so on. Reduced models that ignore particular factors are formed by dropping all X variables involved in those factors from the model.

Fitting the full model yields a scaled deviance value of 69.0 on 64 degrees of freedom. Since 69.0 is well below the upper 5% point of the χ^2_{64} distribution, the five factors appear to explain adequately the variation in proportions of males reporting heart trouble. However, a pertinent question is whether all five

Table 6.1 *Number of subjects sampled (N) and reporting heart trouble (Y), categorized by five explanatory factors. See text for factor level codes.*

SC	Smoke	Alc	Hist Age	1	2	3	1	2	3
1	1	1	Y	3	1	8	0	3	0
			N	81	75	44	36	31	11
		2	Y	0	2	4	2	0	0
			N	140	41	21	34	19	7
	2	1	Y	0	3	20	0	1	6
			N	41	50	85	7	28	21
		2	Y	3	3	12	1	3	5
			N	77	68	65	21	41	16
	3	1	Y	0	3	3	0	2	1
			N	30	26	23	16	15	6
		2	Y	3	2	3	0	3	1
			N	96	54	25	24	34	5
2	1	1	Y	2	2	11	1	0	1
			N	110	53	53	26	15	12
		2	Y	1	2	3	0	1	1
			N	165	54	17	40	16	1
	2	1	Y	0	10	27	0	3	11
			N	34	90	151	13	39	36
		2	Y	1	5	27	1	6	7
			N	72	72	93	15	41	22
	3	1	Y	1	4	13	0	4	6
			N	78	55	96	22	30	12
		2	Y	2	13	14	2	7	3
			N	225	142	68	58	44	9

factors are important, or whether we can find a more parsimonious explanation by using fewer terms in the model. Dropping each factor out of the model in turn, and fitting all possible four-factor linear logistic regression models to these binomial data, yielded the following scaled deviances and residual degrees of freedom:

factors fitted	scaled deviance	residual df
Smoke, Alc, Hist, Age	71.4	65
SC, Alc, Hist, Age	76.8	66
Smoke, SC, Hist, Age	70.1	65
Smoke, SC, Alc, Age	76.7	65
Smoke, SC, Alc, Hist	283.0	66

We see that dropping either social class or alcohol seems to have negligible effect on the scaled deviance (differences of deviance equal to 2.4 and 1.1 respectively, each on 1 degree of freedom – clearly not significant). However, dropping smoking, history or (especially) age leads to a significant increase in scaled deviance, so these factors appear to be important.

Thus we look at the model which includes just smoking, age and history. The scaled deviance turns out to be 72.3 with 66 degrees of freedom. This has a scaled deviance difference from the full model of 3.3 on 2 degrees of freedom, which is evidently not significant. It thus appears that this is our best model. In terms of our dummy variables it includes X_1 (overall mean), X_3, X_4 (Smoke), X_6 (Hist), and X_7, X_8 (Age). Parameter estimates and standard errors for this model are:

Parameter	Estimate	Standard Error
β_1	−4.5	0.24
β_3	0.53	0.18
β_4	0.40	0.19
β_6	0.40	0.15
β_7	1.5	0.24
β_8	2.7	0.23

These estimates show that there is an increased risk of heart trouble for either ex-smokers or current smokers relative to non-smokers (significant positive coefficients of X_3 and X_4), also for those with a family history of heart trouble and for those in the older age ranges.

An interesting feature of these results is the near-equality of the estimates of β_3 and β_4 (certainly the difference between them does not appear to be significant). Thus a simpler model could be tried, in which the groups for smoking are condensed into two: the non-smokers as before, but now combining the ex-smokers and current smokers into a single group of 'smokers'. This is equivalent to setting $\beta_3 = \beta_4$ in the above model, so is nested within it. To fit the new model we remove the previous dummy variables X_3 and X_4, and replace them by a single dummy variable X_9, say, that takes value 0 for non-smokers and 1 for smokers (i.e. $X_9 = X_3 + X_4$). Refitting the model now gives a scaled deviance value of 73.0 on 67 degrees of freedom. This is barely higher than the previous value (a difference of 0.7 on 1 degree of freedom), and only 4.0 higher than the full model for 3 extra degrees of freedom. On either count the change is not significant, so this appears to be the most parsimonious explanatory model. Parameter estimates and standard errors for this model are:

Parameter	Estimate	Standard Error
β_1	−4.5	0.24
β_6	0.41	0.15
β_7	1.5	0.24
β_8	2.7	0.23
β_9	0.47	0.17

Table 6.2 *Standardized deviance residuals for data in Table 6.1.*

SC	Smoke	Alc	Hist Age	1			2		
				1	2	3	1	2	3
1	1	1		1.85	−1.66	0.83	−1.10	0.66	−2.28
		2		−1.81	0.11	0.66	1.57	−1.66	−1.79
	2	1		−1.21	−0.32	0.65	−0.60	−1.37	0.03
		2		1.32	−0.96	−0.49	0.59	−0.69	0.27
	3	1		−1.03	0.83	−0.98	−0.92	0.38	−0.67
		2		1.01	−1.10	−1.18	−1.13	−0.30	−0.43
2	1	1		0.73	−0.29	1.47	0.79	−1.47	−1.14
		2		−0.66	−0.32	0.43	−1.16	−0.07	1.81
	2	1		−1.10	1.46	−1.01	−0.82	−0.58	0.33
		2		−0.22	−0.05	2.01	0.86	0.91	0.38
	3	1		−0.31	0.05	−1.98	−1.08	0.55	1.63
		2		−1.15	1.00	−0.05	0.44	1.20	0.34

To check on the fit of the model we can calculate any of the diagnostics described in Section 5.7. Table 6.2 displays the standardized deviance residuals, and Table 6.3 shows the observed and fitted percentages exhibiting heart trouble for each combination of explanatory factors. A histogram of the residuals resembles a normal curve, and only 2 of the 72 values lie outside the interval $(-2.0, 2.0)$. By and large the comparison of fitted and observed percentages looks reasonable, the big differences generally occurring where observed values are 0% or 100% and usually where the numbers at risk are small. Overall, therefore, we conclude that the fitted model provides a good explanation of the observed data.

A priori structure

A set of comparisons of particular interest when the data fall into *a priori* sets or classes is that of comparing linear predictors between sets. Specifically, if there is just one explanatory variable X, the linear predictor for observation i in set j would typically be

$$\eta_{ij} = \beta_{0j} + \beta_{1j} x_{ij}$$

which is of course just the equation of a straight line. A nested sequence of models of particular interest is then

M_1: separate lines in each set (i.e. β_{0j}, β_{1j} arbitrary);
M_2: common slopes in each set (i.e. β_{0j} arbitrary, $\beta_{1j} = \beta_1$ for all j);
M_3: one common line in all sets (i.e. $\beta_{0j} = \beta_0, \beta_{1j} = \beta_1$ for all j).

Since $M_1 \supset M_2 \supset M_3$, this situation falls within the more general framework outlined earlier. It is also the generalized linear model extension of comparison

Table 6.3 *Observed (O) and fitted (F) percentages reporting heart trouble for each combination of explanatory factors.*

SC	Smoke	Alc	Hist	1			2		
			Age	1	2	3	1	2	3
1	1	1	O	3.7	1.3	18.2	0.0	9.68	0.0
			F	1.1	4.5	14.0	1.6	6.7	19.7
		2	O	0.0	4.9	19.1	5.9	0.0	0.0
			F	1.1	4.5	14.0	1.6	6.7	19.7
	2	1	O	0.0	6.0	23.5	0.0	3.6	28.6
			F	1.7	7.1	20.8	2.5	10.3	28.3
		2	O	3.9	4.4	18.5	4.8	7.3	31.3
			F	1.7	7.1	20.8	2.5	10.3	28.3
	3	1	O	0.0	11.5	13.0	0.0	13.3	16.7
			F	1.7	7.1	20.8	2.5	10.3	28.3
		2	O	3.1	3.7	12.0	0.0	8.8	20.0
			F	1.7	7.1	20.8	2.5	10.3	28.3
2	1	1	O	1.8	3.8	20.8	3.9	0.0	8.3
			F	1.1	4.5	14.0	1.6	6.7	19.7
		2	O	0.6	3.7	17.7	0.0	6.3	100
			F	1.1	4.5	14.0	1.6	6.7	19.7
	2	1	O	0.0	11.1	17.9	0.0	7.7	30.6
			F	1.7	7.1	20.8	2.5	10.3	28.3
		2	O	1.4	6.9	29.0	6.7	14.6	31.8
			F	1.7	7.1	20.8	2.5	10.3	28.3
	3	1	O	1.3	7.3	13.5	0.0	13.3	50.0
			F	1.7	7.1	20.8	2.5	10.3	28.3
		2	O	0.9	9.2	20.6	3.5	15.9	33.3
			F	1.7	7.1	20.8	2.5	10.3	28.3

of linear regressions as discussed in Section 3.6. The method of analysis simply blends the two approaches: we use the dummy variable ideas in Section 3.6 to specify the linear predictors, and then the generalized linear model theory given above for fitting the models and testing the significance of passing between models M_1, M_2 and M_3. This situation is briefly illustrated in the next example.

Example 6.2. In a trial of three insecticides, batches of about fifty insects were exposed to varying deposits of each insecticide; the proportions of insects killed after exposure for six days were as follows (Source: Hewlett and Plackett, 1950).

Insecticide	Amount of Deposit (mg)					
	2.00	2.64	3.48	4.59	6.06	8.00
A	3/50	5/49	19/47	19/38	24/49	35/50
B	2/50	14/49	20/50	27/50	41/50	40/50
C	28/50	37/50	46/50	48/50	48/50	50/50

Taking the logarithm of deposit as the explanatory variable X, we consider linear logistic models for the results with the three insecticides.

Fitting a separate logistic regression to the results of each insecticide is equivalent to having a separate straight line ($\eta_{ij} = \beta_{0j} + \beta_{1j}x_{ij}$ for $i = 1, \ldots, 6$ and $j = A, B, C$) for each linear predictor. This gives total scaled deviance 17.89 on 12 degrees of freedom (18 proportions less 3×2 parameters).

Constraining the slopes of the three straight lines to be the same (i.e. setting $\beta_{1A} = \beta_{1B} = \beta_{1C} = \beta_1$), and hence fitting parallel lines for the linear predictors, gives scaled deviance 21.28 on 14 degrees of freedom. This is an increase in the deviance of 3.39 on 2 degrees of freedom, which is not significant. Hence the parallel lines model with 4 parameters is equivalent to the separate lines with 6 parameters and so is a preferable model.

Finally, constraining the slopes *and* intercepts of the three straight lines to be equal (i.e. additionally setting $\beta_{0A} = \beta_{0B} = \beta_{0C} = \beta_0$), and hence fitting a common line for the three linear predictors, gives scaled deviance 246.83 on 16 degrees of freedom. This is a huge increase of 225.55 on 2 degrees of freedom, and is highly significant. We thus conclude that there are clear differences in intercept between the three linear predictors (which is evident simply from looking at the different overall levels in proportions killed for the three insecticides), but the relationship between proportion killed and (log) dose applied appears to be constant across the three insecticides. A scaled deviance of 21.28 is not significantly high for 14 degrees of freedom, so the parallel lines model is the most parsimonious one for the given data.

Note that a much fuller analysis of this set of data is presented in *Modelling Binary Data* by D. Collett (1991, Chapman & Hall, London), so this reference can be consulted for further details.

6.3.4 Ungrouped binary data

When the data are ungrouped, then each individual in effect forms its own group so that $g = n$. The ith individual has its own set of explanatory variable values x_i and its response variable Y_i takes one of the two possible values 0 or 1. As stated at the beginning of Section 6.2, this situation can be handled by the foregoing methods on setting $N_i = 1$ for all i. However, there is one peculiar aspect of the scaled deviance in this situation that should be noted. Not only does it share the feature of all generalized linear models that its asymptotic null distribution may not be a good approximation to the true null distribution but, more disturbingly, it cannot actually be used as a summary measure of goodness of fit of the model. Why this is so can be seen from the following argument.

The basic set-up here is of n binary variables Y_i that have independent Bernoulli distributions with parameters $\pi_i = \Pr(Y_i = 1)$ and logistic link function

$$\log\left(\frac{\pi_i}{1 - \pi_i}\right) = \beta_1 x_{i1} + \cdots + \beta_p x_{ip} \tag{6.2}$$

for $i = 1, \ldots, n$. The means of the Bernoulli variables are $\mu_i = \pi_i$ and the log-likelihood of the data is

$$l(\beta_1, \ldots, \beta_p) = \sum_{i=1}^{n} \{Y_i \log \pi_i + (1 - Y_i) \log(1 - \pi_i)\}. \tag{6.3}$$

Now suppose that we have fitted the model by maximum likelihood and obtained estimates $\hat{\beta}_1, \ldots, \hat{\beta}_p$. Then substituting $N_i = 1$ successively in equation (6.1) and in the scaled deviance below it yields the scaled deviance of the present model in the expanded form

$$D = 2 \sum_{i=1}^{n} [Y_i \log Y_i - Y_i \log \hat{\pi}_i + (1 - Y_i) \log(1 - Y_i) - (1 - Y_i) \log(1 - \hat{\pi}_i)]. \tag{6.4}$$

But both $Y_i \log Y_i$ and $(1 - Y_i) \log(1 - Y_i)$ equal zero for each of the only two values $(0, 1)$ that Y_i can take, so the scaled deviance reduces to

$$D = -2 \sum_{i=1}^{n} [Y_i \log \hat{\pi}_i + (1 - Y_i) \log(1 - \hat{\pi}_i)]$$

$$= -2 \sum_{i=1}^{n} [Y_i \log\{\hat{\pi}_i / (1 - \hat{\pi}_i)\} + \log(1 - \hat{\pi}_i)]. \tag{6.5}$$

But the maximum likelihood estimates have been obtained as the solutions of the equations

$$\frac{\partial l}{\partial \beta_j} = 0 \qquad \text{for } j = 1, \ldots, p,$$

and from the model specification we obtain

$$\frac{\partial l}{\partial \beta_j} = \sum_{i=1}^{n} \left\{ \frac{Y_i}{\pi_i} - \frac{1 - Y_i}{1 - \pi_i} \right\} \pi_i (1 - \pi_i) x_{ij}$$

$$= \sum_{i=1}^{n} (Y_i - \pi_i) x_{ij}. \tag{6.6}$$

Multiplying this equation by β_j and summing over j yields

$$\sum_{j=1}^{p} \beta_j \frac{\partial l}{\partial \beta_j} = \sum_{i=1}^{n} (Y_i - \pi_i) \sum_{j=1}^{p} \beta_j x_{ij} = \sum_{i=1}^{n} (Y_i - \pi_i) \log[\pi_i / (1 - \pi_i)] \tag{6.7}$$

on substituting from the definition of the link function in (6.2). But the left-hand side of (6.6), and hence that of (6.7), must be zero at the maximum likelihood estimators of the β_j, from which it follows that

$$Y_i \log[\hat{\pi}_i/(1 - \hat{\pi}_i)] = \hat{\pi}_i \log[\hat{\pi}_i/(1 - \hat{\pi}_i)].$$

Substituting this result into equation (6.5) reduces the scaled deviance of the model to

$$D = -2 \sum_{i=1}^{n} [\hat{\pi}_i \log\{\hat{\pi}_i/(1 - \hat{\pi}_i)\} + \log(1 - \hat{\pi}_i)], \qquad (6.8)$$

which depends only on the fitted values $\hat{\pi}_i$ and *not on the observed* Y_i. Thus the scaled deviance can tell us nothing about the correspondence between observed and expected values, i.e. about the goodness of fit of the model.

Of course, the *difference* between scaled deviances of two nested models can be used as before when investigating models to fit to such data.

Example 6.3. To study the effect of volume (X_1) and rate (X_2) of air inspired by human subjects on the occurrence or not (Y, 1 or 0) of transient vaso-constriction response in the skin of the fingers, 39 observations on these variables were obtained from 3 subjects in a laboratory. The data are presented in the first four columns of Table 6.4.

Assuming independence of observations (including those on the same subject), a linear logistic regression with Bernoulli variables (i.e. binomial with $N_i = 1$) was fitted to these data. Specifying the linear predictor as

$$\eta_i = \beta_0 + \beta_1 x_{i1} + \beta_2 x_{i2},$$

parameter estimates and standard errors were given by

Parameter	Estimate	Standard Error
β_0	−9.530	3.224
β_1	2.649	0.912
β_2	3.882	1.425

The scaled deviance of 29.77 on 36 degrees of freedom is of no interest for the reasons given above. However, all parameters appear to be significant from approximate t-tests. Fitted values and standardized Pearson residuals are given in columns 5 and 6 of Table 6.4. The sum of squares of these residuals gives a Pearson X^2 value of 38.85 on 36 degrees of freedom, which suggests a good fit of model to data. However, residuals for observations 4 and 18 seem large so should be investigated to see if there is some error in the recorded values.

Table 6.4 *Vaso-constriction data, fitted values and standardized Pearson residuals (Source: Finney, 1947).*

i	x_{i1}	x_{i2}	Y_i	$\hat{\pi}_i$	$r_{i(P)}$
1	3.70	0.83	1	0.999	0.030
2	3.50	1.09	1	0.999	0.031
3	1.25	2.50	1	0.875	0.378
4	0.75	1.50	1	0.066	3.751
5	0.80	3.20	1	0.886	0.358
6	0.70	3.50	1	0.921	0.292
7	0.60	0.75	0	0.005	−0.074
8	1.10	1.70	0	0.320	−0.685
9	0.90	0.75	0	0.017	−0.132
10	0.90	0.45	0	0.008	−0.089
11	0.80	0.57	0	0.007	−0.086
12	0.55	2.75	0	0.473	−0.947
13	0.60	3.00	0	0.679	−1.453
14	1.40	2.33	1	0.889	0.354
15	0.75	3.75	1	0.965	0.191
16	2.30	1.64	1	0.977	0.154
17	3.20	1.60	1	0.999	0.028
18	0.85	1.42	1	0.077	3.458
19	1.70	1.06	0	0.470	−0.941
20	1.80	1.80	1	0.903	0.328
21	0.40	2.00	0	0.064	−0.262
22	0.95	1.36	0	0.096	−0.326
23	1.35	1.35	0	0.329	−0.700
24	1.50	1.36	0	0.474	−0.950
25	1.60	1.78	1	0.802	0.497
26	0.60	1.50	0	0.038	−0.199
27	1.80	1.50	1	0.807	0.489
28	0.95	1.90	0	0.308	−0.668
29	1.90	0.95	1	0.590	0.834
30	1.60	0.40	0	0.095	−0.323
31	2.70	0.75	1	0.950	0.230
32	2.35	0.03	0	0.419	−0.849
33	1.10	1.83	0	0.399	−0.814
34	1.10	2.20	1	0.638	0.752
35	1.20	2.00	1	0.605	0.808
36	0.80	3.33	1	0.917	0.302
37	0.95	1.90	0	0.308	−0.668
38	0.75	1.90	0	0.170	−0.453
39	1.30	1.63	1	0.456	1.093

6.4 Related methods

Although the most common link function for either binary or grouped binary data is the logistic link, the user is of course at liberty to choose any other link function that preserves the right features of the data (in particular that satisfies the necessary constraints regarding probabilities and their ranges). Other features of the generalized linear model (i.e. the linear predictor, the distributional assumption and the definition of the scaled deviance in terms of observed and fitted values) will remain unchanged, but changing the link function will necessarily change the estimated parameters of the linear predictor and hence also the value of the scaled deviance. Thus it may happen that different link functions are best for different sets of data, even though all sets might share some general overall features.

In this section, we briefly consider other possible link functions that might be used. The two main contenders are the *probit* link and the *complementary log–log* link; we first consider the necessary changes to the iterative maximum likelihood estimation scheme produced by each of these link functions, then we indicate the sort of situations for which they are particularly suitable, mentioning any specific results that are useful in these situations, and finally we compare the result of all three methods on the data of Example 6.3.

6.4.1 Amendments to iterative schemes

The mathematics given in Section 6.3.2, providing details of the iterative maximum likelihood algorithm, rested on the fact that the link function was the logistic. We thus need to produce the correponding calculations for each of the other link functions that might be used. We retain the same linear predictor definition as before, viz. $\eta_i = \beta_1 x_{i1} + \cdots + \beta_p x_{ip}$, and once again consider the grouped binary case for most generality. Thus $\pi_i = \Pr(Y_i = 1)$ and $\mu_i = N_i \pi_i$ as before.

The probit link

This is defined by

$$\eta_i = \Phi^{-1}(\pi_i),$$

or in other words

$$\pi_i = \frac{\mu_i}{N_i} = \int_{-\infty}^{\eta_i} \phi(u)\, du$$

where $\phi(u)$ is the standard normal density function $\frac{1}{\sqrt{2\pi}} \exp(-\frac{1}{2} u^2)$. Thus, directly from the definition of integration as the inverse of differentiation, it follows that

$$\frac{1}{N_i} \frac{\partial \mu_i}{\partial \eta_i} = \phi(\eta_i)$$

so that

$$\frac{\partial \mu_i}{\partial \eta_i} = N_i \phi(\eta_i).$$

Since the distributional assumptions remain the same as before, i.e. that $Y_i \sim$ binomial(N_i, π_i), then $\text{Var}(Y_i) = N_i \pi_i (1 - \pi_i)$ as before. Thus given 'current' values $\tilde{\beta}_j$ for all the parameters β_j, successive substitution yields 'current' values $\tilde{\eta}_i, \tilde{\pi}_i, \tilde{\mu}_i$ and hence those of $\text{Var}(Y_i)$ and $\frac{\partial \mu_i}{\partial \eta_i}$. The latter requires either numerical integration of the standard normal density function or access to detailed tabulation of the standard normal integral, but appropriate routines in either case are now available on most computer installations. Thus all necessary quantities needed for the iterative scheme, as well as for obtaining standard errors of estimates, can be found readily.

The complementary log–log link

In this case we have

$$\eta_i = \log[-\log(1 - \pi_i)],$$

which leads to

$$\pi_i = \frac{\mu_i}{N_i} = 1 - \exp\{-e^{\eta_i}\}$$

and hence to

$$\frac{\partial \mu_i}{\partial \eta_i} = N_i \exp\{\eta_i - e^{\eta_i}\}.$$

Once again $\text{Var}(Y_i) = N_i \pi_i (1 - \pi_i)$, so given 'current' values $\tilde{\beta}_j$ successive substitution yields 'current' values $\tilde{\eta}_i, \tilde{\pi}_i, \tilde{\mu}_i$ and hence those of $\text{Var}(Y_i)$ and $\frac{\partial \mu_i}{\partial \eta_i}$. Thus all necessary quantities are again available for the iterative scheme to proceed and for standard errors to be calculated.

6.4.2 Bioassay

Historically, one of the first uses of regression models for binomial data was for bioassay results. Responses were typically the proportions or percentages of animals or insects reacting to various dose levels of a toxic substance, and the original name for such data was *quantal responses*. Typically the only explanatory variable in such studies was the dose of toxin administered, and furthermore it was usually assumed that the probability of response of a single insect depended on the logarithm of the applied dose. Denoting this log dose in the ith group of insects by x_i and the probability of response by π_i, the linear predictor was traditionally thus taken to be $\eta_i = \beta_0 + \beta_1 x_i$.

Background theory suggested that there existed a *tolerance distribution* $f(u)$ such that the probability π of response by an insect in the ith group depended on the cumulative distribution function of the tolerance distribution evaluated at the linear predictor for that group. Taking the tolerance distribution to be the normal distribution (the assumption made in the early days of this work) leads to the model of this chapter with *probit* link function. In this case, therefore, we have

$$\pi_i = \Phi(\beta_0 + \beta_1 x_i).$$

More sophisticated studies that followed assumed an *extreme value distribution* for the tolerance distribution, and this produces the model of this chapter with the *complementary log–log* link. Here we have

$$\pi_i = 1 - \exp\{-e^{\beta_0 + \beta_1 x_i}\}.$$

Nowadays, for simplicity and convenience, the *logistic* distribution would be usually assumed for the tolerance distribution. This leads to the model of this chapter with *logistic* link and

$$\pi_i = \frac{e^{\beta_0 + \beta_1 x_i}}{1 + e^{\beta_0 + \beta_1 x_i}}.$$

Interest in bioassay often focusses on what is known as an *effective dose*, namely the amount of chemical required to produce a response in a given percentage of the individuals exposed to it. Although any arbitrary percentage may be specified, the two commonly used values are 50% and 90%. Thus we denote by $ED50$ the amount of chemical required to produce a response in 50% of individuals, and by $ED90$ the amount of chemical required to produce a response in 90% of individuals. If the response is death, then the word *lethal* often replaces *effective*, so that these quantities might be referred to as $LD50$ and $LD90$ respectively.

To relate these quantities to the model parameters, we make the connection that $c\%$ of the individuals show a response if $\Pr(response) = c/100$. Thus the $ED50$ value of X is that value $x_{0.5}$ for which $\pi_i = 0.5$, the $ED90$ value is that value $x_{0.9}$ for which $\pi_i = 0.9$, and so on. For all the possible link functions specified above, therefore, any of these values x_α is given as the solution of an equation of the form

$$\beta_0 + \beta_1 x_\alpha = k_\alpha,$$

where k_α depends on the specified percentage c and on the chosen link function. Thus the required solution is

$$x_\alpha = \frac{k_\alpha - \beta_0}{\beta_1}. \tag{6.9}$$

In particular, for the $ED50$ we have $c = 50$, $\alpha = 0.5$ and $\pi_i = 0.5$ so that $k_{0.5} = 0$ for both probit and logistic link functions and $x_{0.5} = \beta_0/\beta_1$ for these link functions. In the case of the complementary log–log link function, however, $k_{0.5} = \log_e(\log_e 2) = -0.367$. For the $ED90$ we have $k_{0.9} = 1.282$ for the probit link (from tables of the cumulative normal probability distribution), $k_{0.9} = \log_e(0.9/0.1) = 2.197$ for the logit link, and $k_{0.9} = \log_e(\log_e 10) = 0.834$ for the complementary log–log link.

When we fit the model to the data, we obtain estimates $\hat{\beta}_0$, $\hat{\beta}_1$ of the parameters along with their approximate variances and the covariance between them. The estimate of the appropriate effective dose is thus given by

$$\hat{x}_\alpha = \frac{k_\alpha - \hat{\beta}_0}{\hat{\beta}_1}. \tag{6.10}$$

We would usually also want to find a confidence interval for x_α, but this seems to present problems as regards underlying theory. Our asymptotic results (Section 5.3) tell us that $\hat{\beta}_0$ and $\hat{\beta}_1$ are both approximately normal and unbiased, so we effectively need to find a confidence interval for a quantity estimated by the ratio of two normal random variables. This can be done using a result known as *Fieller's Theorem*, the general argument of which runs as follows.

Suppose that $\rho = \gamma_1/\gamma_2$, where γ_1, γ_2 are estimated by $\hat{\gamma}_1$, $\hat{\gamma}_2$ and these estimates are normally distributed with means γ_1, γ_2, variances v_1, v_2 and covariance v_{12}. Consider $\hat{\psi} = \hat{\gamma}_1 - \rho\hat{\gamma}_2$. Then

$$E(\hat{\psi}) = \gamma_1 - \rho\gamma_2 = 0,$$

and

$$\mathrm{Var}(\hat{\psi}) = V = v_1 + \rho^2 v_2 - 2v_{12}\rho.$$

Since both $\hat{\gamma}_1$ and $\hat{\gamma}_2$ are normally distributed then so is $\hat{\psi}$, and hence

$$\frac{\hat{\gamma}_1 - \rho\hat{\gamma}_2}{\sqrt{V}}$$

has a standard normal distribution. Thus a $100(1 - \alpha)\%$ confidence interval for ρ is the set of values for which

$$|\hat{\gamma}_1 - \rho\hat{\gamma}_2| \leq z_{\alpha/2}\sqrt{V}$$

where $z_{\alpha/2}$ is the upper $\alpha/2$ point of the standard normal distribution. Squaring both sides, taking the equality, substituting for V from above and tidying up the algebra yields the following quadratic equation in ρ:

$$(\hat{\gamma}_2^2 - v_2 z_{\alpha/2}^2)\rho^2 + (2v_{12}z_{\alpha/2}^2 - 2\hat{\gamma}_1\hat{\gamma}_2)\rho + \hat{\gamma}_1^2 - v_1 z_{\alpha/2}^2 = 0. \tag{6.11}$$

The two roots of this equation constitute the confidence limits for ρ.

To apply this result to our bioassay problem, simply use $k_\alpha - \hat{\beta}_0$ for $\hat{\gamma}_1$, $\hat{\beta}_1$ for $\hat{\gamma}_2$, and replace v_1, v_2 and v_{12} by the appropriate elements of the estimated variance–covariance matrix of $\hat{\beta}_0$, $\hat{\beta}_1$. This will give confidence limits for x_α. If the logarithm of dose has been used for X, then exponentiating these limits will give appropriate limits for the dose itself.

Example 6.4. We return to the vaso-constriction response data of Example 6.3 and Table 6.4, to illustrate some similarities and differences when fitting the various link functions discussed above. If instead of the logit link of Example 6.3 we fit a probit link (retaining the same linear predictor), we obtain the following parameter estimates

Parameter	Estimate	Standard Error
β_0	−5.192	1.592
β_1	1.476	0.461
β_2	2.117	0.716

Fitted values and standardized Pearson residuals are given in columns 2 and 3 of Table 6.5; the resulting Pearson X^2-statistic value is 34.54 on 36 degrees of freedom. Thus the parameter estimates and their standard errors differ appreciably from those with the logit link, but the fit is slightly better. Note also that observations 4 and 18 still show high residuals.

Using the complementary log–log link gives estimates much more like the logit ones:

Parameter	Estimate	Standard Error
β_0	−8.329	2.572
β_1	2.127	0.652
β_2	3.294	1.132

However, some convergence problems were encountered in fitting this model, with some fitted values having to be constrained to equal their limits. The fitted values and standardized Pearson residuals are shown in columns 4 and 5 of Table 6.5 (where residuals of zero indicate that the corresponding units were constrained to equal their limiting values). The Pearson X^2-statistic value is 37.10 on 36 degrees of freedom, so the fit is very comparable to that with logit link. Note that observations 4 and 18 are still poorly fitted, perhaps due to misclassification of Y.

6.5 Ordered polytomous data

In many practical situations, the response variable may be graded into a number of categories rather than being just 'yes' or 'no'. For example, an individual may be deemed to have either a 'mild' or an 'average' or a 'severe' form of a condition; the similarity of two soft drinks may be rated on a scale running from 1 to 5 by a tasting panel; or a questionnaire may contain a number of statements and respondents might be asked to assess whether they 'strongly agree with', 'agree with', 'are neutral to', 'disagree with' or 'strongly disagree with' each statement. In general, such situations can be characterized by a response variable Y which takes values 0 (denoting absence of response) or $1, 2 \ldots, k$ (denoting the categories of response in increasing order). Moreover, if the data are again grouped into g groups by values of the explanatory variables, then N_i denotes the number of respondents in the ith group and Y_{ij} denotes the number in this group that respond in the jth category (i.e. Y_{ij} is the number for which $Y_i = j$).

Table 6.5 *Vaso-constriction data, probit and complementary log–log links.*

	Probit		Comp log–log	
i	$\hat{\pi}_i$	$r_{i(P)}$	$\hat{\pi}_i$	$r_{i(P)}$
1	1.000	0.008	1.000	0.000
2	1.000	0.008	1.000	0.000
3	0.874	0.380	0.952	0.226
4	0.082	3.343	0.067	3.729
5	0.890	0.352	0.953	0.223
6	0.927	0.280	0.984	0.127
7	0.002	−0.049	0.009	−0.093
8	0.361	−0.752	0.286	−0.633
9	0.015	−0.122	0.023	−0.153
10	0.004	−0.066	0.012	−0.111
11	0.004	−0.063	0.011	−0.107
12	0.512	−1.025	0.402	−0.819
13	0.693	−1.504	0.643	−1.342
14	0.887	0.357	0.968	0.181
15	0.973	0.166	1.000	0.016
16	0.982	0.135	1.000	0.000
17	1.000	0.006	1.000	0.000
18	0.096	3.069	0.077	3.452
19	0.488	−0.977	0.464	−0.930
20	0.899	0.366	0.985	0.125
21	0.082	−0.298	0.062	−0.256
22	0.120	−0.370	0.095	−0.324
23	0.366	−0760	0.305	−0.663
24	0.496	−0.992	0.457	−0.917
25	0.794	0.509	0.874	0.379
26	0.044	−0.214	0.041	−0.208
27	0.797	0.504	0.890	0.351
28	0.353	−0.739	0.270	−0.608
29	0.592	0.831	0.614	0.793
30	0.112	−0.355	0.104	−0.341
31	0.948	0.233	1.000	0.013
32	0.431	−0.871	0.447	−0.899
33	0.435	−0.878	0.359	−0.748
34	0.649	0.735	0.623	0.778
35	0.618	0.787	0.588	0.837
36	0.922	0.292	0.982	0.135
37	0.353	−0.739	0.270	−0.608
38	0.212	−0.518	0.150	−0.420
39	0.483	1.034	0.426	1.161

Strictly, this situation lies beyond the scope of the present chapter because the responses are no longer binary and the binomial distribution is no longer the correct basis for analysis. However, we mention the situation briefly here because it is a natural extension of the earlier ones and also because the models we have met previously continue to be appropriate in this case also.

We could write $\pi_{ij} = \Pr(Y_i = j)$ and formulate models as before for these category probabilities. However, it turns out that simple models for the *cumulative* response probabilities $\gamma_{ij} = \Pr(Y_i \leq j)$ have better properties. In particular, simple linear predictors may be linked to these cumulative probabilities via the logistic ($\log\{\gamma_{ij}/[1 - \gamma_{ij}]\}$), probit ($\Phi^{-1}[\gamma_{ij}]$), or complementary log–log ($\log\{-\log[1 - \gamma_{ij}]\}$) functions very satisfactorily. This is the same as the binary case treated earlier in the chapter, but for the distribution of the data we now need to generalize the binomial to the *multinomial* distribution (Section 2.3.1). This leads to different likelihoods and hence to different quantities in the iterative maximum likelihood estimation scheme. We do not give any further details here, but refer the interested reader to *An Introduction to Categorical Data Analysis* by A. Agresti (Wiley, 1996, Chapter 8) for a more extensive discussion and illustrative example.

7
Tables of counts and log–linear models

7.1 Introduction

Probably the most commonly encountered type of data in many areas of application (particularly in the social sciences) is a count of the numbers of individuals who either possess certain combinations of attributes or who can be classified according to certain combinations of factors, presented in a *contingency* table. The following three data sets give examples of such tables, in increasing order of complexity. These, and other similar data sets, are described in more detail by S.E. Fienberg in *The Analysis of Cross-classified Categorical Data* (1980, 2nd edition, The MIT Press, Cambridge, Massachusetts), but we here give just their salient features.

Table 7.1 shows 205 married couples, classified according to the heights of each partner (thereby yielding a simple *two-way* contingency table). A question of interest here might be whether there is any dependency between husband's and wife's height in the population of all married couples.

Table 7.2 shows the numbers of babies who died and the numbers who survived in each of two clinics, additionally classified according to the level of antenatal care received. Since there are now three classifying factors this is a *three-way* table.

There are a few general concepts connected with such higher-way tables that can conveniently be stated here. Summing over the levels of any one factor produces a two-way *marginal* table (for example, the values in the 'totals' column of Table 7.2 define the 'clinic' × 'care' two-way marginal table) while summing over the levels of any two factors produces a one-way, i.e. single-factor, marginal table (for example, the values 689 and 26 in the 'totals' row of Table 7.2 define the one-way marginal table for the factor 'survival'). We can also obtain *conditional* tables, by taking values over all other factors for any level of a single factor or combination of levels for two factors. For example, the values in the 'survived' column of Table 7.2 yield the conditional 'clinic' × 'care' two-way table for the surviving babies.

Returning to Table 7.2 itself, obvious questions to ask are whether the chances of survival for a single baby depend on the factors 'clinic' and 'antenatal care',

Table 7.1 *205 married couples, classified according to the heights of each partner (Source: Yule, 1900).*

Husband	Wife Tall	Wife Medium	Wife Short	Totals
Tall	18	28	14	60
Medium	20	51	28	99
Short	12	25	9	46
Totals	50	104	51	205

and whether there is any evidence of an interaction between these factors (i.e. whether the chance of survival varies differently between clinics for each level of antenatal care).

Table 7.3 shows data on coal miners between the ages of 20 and 64, classified by age group, whether or not they suffered from breathlessness, and whether or not they suffered from wheeze. Questions of interest here include whether the prevalence of each condition varies with age, whether the two conditions are associated, and whether this association varies with age.

It can be seen that the structure of all these data sets is very reminiscent of those in Chapter 4, with interest focussing on the effect of various *factors* (such as 'clinic', 'breathlessness', 'age', etc) on the response in the body of the table. Indeed, we can readily phrase our questions about the data in the terminology of that chapter: is there a *main effect* of 'clinic'? Is there an *interaction* between 'breathlessness' and 'age'? etc. However, the form of the data prevents us from answering these questions by means of the analysis of variance technique described in Chapter 4, because this technique requires our response data to be *continuous* and, preferably, *normally distributed* but in the present case all our responses are *counts* and hence *discrete*. Note that earlier, in Section 3.5.4, we discussed the use of transformations of the data in those situations where the assumptions necessary for a particular technique were not satisfied, in the hope that the transformed data would be more amenable to the

Table 7.2 *715 babies classified according to the clinic in which they were born, the level of antenatal care they received, and whether or not they survived (Source: Bishop, 1969).*

Clinic	Antenatal Care	Survived	Died	Totals
A	Low	176	3	179
	High	293	4	297
B	Low	197	17	214
	High	23	2	25
	Totals	689	26	715

Table 7.3 *Coal miners classified by age, breathlessness and wheeze (Source: Ashford and Sowden, 1970).*

Age group in years	Breathlessness Wheeze	Breathlessness No wheeze	No breathlessness Wheeze	No breathlessness No wheeze	Totals
20 – 24	9	7	95	1841	1952
25 – 29	23	9	105	1654	1791
30 – 34	54	19	177	1863	2113
35 – 39	121	48	257	2357	2783
40 – 44	169	54	273	1778	2274
45 – 49	269	88	324	1712	2393
50 – 54	404	117	245	1324	2090
55 – 59	406	152	225	967	1750
60 – 64	372	106	132	526	1136
Totals	1827	600	1833	14,022	18,282

technique. It turns out that when our responses are counts we can approximately satisfy analysis of variance assumptions by taking square roots of the responses and analysing the transformed instead of the raw values; this is the procedure that used to be employed for such data, but the resulting analysis was necessarily only approximate.

However, with the development of generalized linear models it is now possible to conduct an exact analysis of such data, and this mode of analysis is therefore described in the present chapter. We first consider ways in which such data can arise in practice and develop the appropriate distributional assumptions for the analysis. We then consider how the concepts inherent in analysis of variance can be translated into those for analysis of contingency tables. Finally, we discuss the details of the analysis, for a range of practically relevant situations.

7.2   Data mechanisms and distributions

Cross-classified contingency table data can be obtained from a number of distinct sampling mechanisms, each of which gives rise to its own specific distributional model. It is important to distinguish these different mechanisms, as they have some implications in the process of analysing the data. It turns out that the mechanisms are equivalent to each other with respect to the basic likelihood, but they can have important differences when it comes to deciding on which terms to include in fitted models. We therefore first describe the three most common types in general terms, and then consider some of the more technical aspects.

7.2.1 Sampling designs

In fact there are many possible designs that produce contingency table data, but we focus on the three most common types. To fix ideas, we illustrate each type by reference to the data on survival of babies given in Table 7.2.

(i) Nothing fixed by design.

This type of design would occur if we simply recorded information on clinic and antenatal care for every baby born during one month. The number of births is thus unknown at the start of the study, so the value 715 here is a realization of a random variable.

(ii) Total sample size fixed.

In this case we decide at the outset that we will continue the study until we have recorded information on 715 babies and we will then stop. The total number of births is thus fixed in advance, but the numbers in each clinic and antenatal care group, along with numbers surviving and dying, are not fixed.

(iii) One or more margins fixed.

Here we decide some more detailed features at the outset. For example, we might decide to record information on antenatal care and survival separately on all babies born at each clinic, stopping when we have 476 from clinic A and 239 from clinic B. In this case the 'clinic' margin is fixed (and, by implication, so is the total sample size). Alternatively, we might decide to record information on survival, stopping when we have 179 from clinic A that received low antenatal care, 297 from clinic A that received high antenatal care, 214 from clinic B that received low antenatal care, and 25 from clinic B that received high antenatal care. In this case the 'clinic' × 'care' two-way margin is fixed in advance (and so, by implication, are the separate 'clinic' and 'care' one-way margins as well as the total sample size).

Any of the above schemes can be adopted however many factors are used to classify the individuals. In particular, if there are m factors and hence an m-way contingency table, then scheme (iii) allows any of the j-way margins to be fixed, for $j < m$. Note, however, that if we fix a particular j-way margin then by implication all margins of the resulting j-way marginal table are also fixed. We return to this point when discussing choice of model later.

7.2.2 Distributions

To discuss distributional models appropriate for each of the above sampling designs, we first need to establish some basic notation. Given the similarity of structure of contingency table data to that of factorial analysis of variance situations, the notation will follow that of Chapter 4 in general terms. However, data comprising counts have their own specific features which need to be highlighted.

We suppose in general that there are m classifying factors and hence the data consist of entries in an m-way contingency table. Denote the classifying factors by $F_1, F_2, \ldots, F_m$, and suppose that they have $I, J, \ldots, K$ levels respectively. The table thus comprises $C = I \times J \times \cdots \times K$ *cells*, one for each possible combination of levels of the factors. The entries in these cells are realizations of independent random variables, the independence resulting from the fact that each entry represents a count (i.e. sum) over individuals in a random sample and such individuals are mutually independent by definition. Let us therefore denote by $Y_{ij\cdots k}$ the random variable in the cell defined by levels $i, j, \ldots, k$ of factors $F_1, F_2, \ldots, F_m$ respectively, let the mean of this random variable be $\mu_{ij\cdots k}$, and let the actual entry in the cell be $y_{ij\cdots k}$. Suppose that the total number of individuals in the sample is n.

A probabilistic framework for contingency table data is achieved by assuming that the data are obtained as a random sample from the population of interest, and that $\pi_{ij\cdots k}$ denotes the probability that any randomly chosen member of the population will exhibit level i of F_1, level j of $F_2, \ldots$, level k of F_m. Then, from independence of entries in the table, the likelihood of the given contingency table is

$$L(y, \rho) = \prod_{i=1}^{I} \prod_{j=1}^{J} \cdots \prod_{k=1}^{K} \pi_{ij\cdots k}^{y_{ij\cdots k}} \tag{7.1}$$

and the log-likelihood is

$$l(y, \rho) = \sum_{i=1}^{I} \sum_{j=1}^{J} \cdots \sum_{k=1}^{K} y_{ij\cdots k} \log \pi_{ij\cdots k}. \tag{7.2}$$

The various sampling designs listed above differ in respect of the model that is appropriate for generating the $\pi_{ij\cdots k}$, so we now consider these models.

Nothing fixed

In this case it is reasonable to suppose that we observe a set of Poisson processes, one for each cell in the table and with no constraint on the total number of observations. Consequently we can assume that the $Y_{ij\cdots k}$ are independent Poisson random variables with means $\mu_{ij\cdots k}$. The probability of observing $y_{ij\cdots k}$ individuals at level i of F_1, level j of $F_2, \ldots$, level k of F_m is thus the Poisson probability (Section 2.3.1) $\dfrac{\mu_{ij\cdots k}^{y_{ij\cdots k}} e^{-\mu_{ij\cdots k}}}{y_{ij\cdots k}!}$, so that the likelihood of the table is

$$L(y, \rho) = \prod_{i=1}^{I} \prod_{j=1}^{J} \cdots \prod_{k=1}^{K} \frac{\mu_{ij\cdots k}^{y_{ij\cdots k}} e^{-\mu_{ij\cdots k}}}{y_{ij\cdots k}!}. \tag{7.3}$$

A well-known feature of the Poisson distribution is that the sum of independent Poisson random variables is also a Poisson variable, whose mean equals the sum of the means of the constituent Poissons. In the present case this result implies that the total·count n in the table is itself a realization of a Poisson variable, with mean $\sum_{i=1}^{I} \sum_{j=1}^{J} \cdots \sum_{k=1}^{K} \mu_{ij \cdots k}$.

Total sample size fixed

Here we take a fixed sample of size n and cross-classify each member of the sample according to its levels of the classifying factors. The appropriate distributional model is thus a *multinomial* one (Section 2.3.1) in which n observations are distributed among C categories, the category probabilities being the $\pi_{ij \cdots k}$. The likelihood of the table is thus

$$L(y, \rho) = \frac{n!}{\prod_{i=1}^{I} \prod_{j=1}^{J} \cdots \prod_{k=1}^{K} y_{ij \cdots k}!} \prod_{i=1}^{I} \prod_{j=1}^{J} \cdots \prod_{k=1}^{K} \pi_{ij \cdots k}^{y_{ij \cdots k}}. \tag{7.4}$$

In this case we have $\mu_{ij \cdots k} = n \pi_{ij \cdots k}$.

Specified margins fixed

Suppose a particular margin is fixed, say the totals $y_{++\cdots+k}$ $(k = 1, \ldots, K)$ obtained by summing the $y_{ij \cdots k}$ over all factors except F_m, for each of the K levels of F_m. Let $\rho_{ij \cdots k}$ denote the conditional probability that an individual which is at level k of F_m is also at levels i of F_1, j of F_2, and so on (so that, by the usual rules of conditional probability, $\rho_{ij \cdots k} = \pi_{ij \cdots k}/\pi_{++\cdots+k}$, where the last term is the marginal probability that an individual is at level k of F_m). Then for the kth level of F_m, we have a fixed sample size $y_{++\cdots+k}$ and the individuals are distributed among the combinations of levels of the other factors in multinomial fashion with category probabilities $\rho_{ij \cdots k}$. Thus the likelihood of the portion of the table corresponding to level k of F_m is

$$\frac{y_{++\cdots+k}!}{\prod_{i=1}^{I} \prod_{j=1}^{J} \cdots y_{ij \cdots k}!} \prod_{i=1}^{I} \prod_{j=1}^{J} \cdots \rho_{ij \cdots k}^{y_{ij \cdots k}}.$$

Independence of the individuals at the different levels of F_m then means that the likelihood of the whole table is given by the product of such terms, one for each level of F_m. This likelihood, known as the *product-multinomial* is thus given by

$$L(y, \rho) = \prod_{k=1}^{K} \left(\frac{y_{++\cdots+k}!}{\prod_{i=1}^{I} \prod_{j=1}^{J} \cdots y_{ij \cdots k}!} \prod_{i=1}^{I} \prod_{j=1}^{J} \cdots \rho_{ij \cdots k}^{y_{ij \cdots k}} \right). \tag{7.5}$$

The same principle is followed whatever the size of margin that is fixed.

7.3   Log–linear models for means

Having considered the distributional, i.e. random, aspects let us now turn to the systematic features of a contingency table. We have seen in other chapters that the systematic features of a model are generally embodied in the means of the associated random variables, so we consider the means of the cells in the table. To introduce the ideas without overcomplicating the notation, it is most convenient to look at the simplest case of a two-way table in detail and then to argue by analogy for extension to higher-way tables.

If we have a two-way table, such as that in Table 7.1 for married couples classified according to heights of each partner, then the notation is somewhat simpler than the general notation of the previous section. The two classifying factors can be denoted by F_1 and F_2, and let us assume that they have I and J levels respectively (so that there are $C = I \times J$ cells in the table). Thus, in Table 7.1, we can arbitrarily assign F_1 to 'husband's height' and F_2 to 'wife's height', and we see that since each has categories 'tall', 'medium' and 'small' then $I = J = 3$.

The entries in the table, and their associated random variables, means and probabilities now only have to be indexed by two factors, so we can denote by Y_{ij} the random variable giving the count in the cell corresponding to levels i of F_1 and j of F_2, by y_{ij} the realization of Y_{ij} in a particular table, by μ_{ij} the mean of Y_{ij}, and by π_{ij} the probability that a random individual will be classified into this cell.

In line with the previous notation, let us denote by a '+' in a subscript the summation over that particular index to form a marginal total or probability. Thus, for example, y_{+j} denotes the marginal total for level j of F_2, y_{++} denotes the total over both factors (i.e. the sample size n), μ_{+j} and μ_{++} denote the means of the random variables corresponding to these two marginal totals, and π_{i+} denotes the marginal probability that a randomly chosen individual falls in level i of F_1 (irrespective of that individual's level of F_2). Referring to Table 7.1 again, therefore, we see that $y_{12} = 28$, $y_{23} = 28$, $y_{+2} = 104$, $y_{3+} = 46$, $y_{++} = 205$ and so on.

The main emphasis in classical analysis of the two-way contingency table has usually been on testing the null hypothesis that there is no association between the two classifying factors, i.e. that F_1 behaves independently of F_2. Under this hypothesis, standard probability theory tells us that

$$\pi_{ij} = \pi_{i+}\pi_{+j}.$$

But since $\mu_{ij} = n\pi_{ij}$ for all i and j, then if there is no association between the two factors we have

$$\mu_{ij} = \frac{\mu_{i+}\mu_{+j}}{n}. \tag{7.6}$$

Taking logarithms in this equation thus yields

$$\log \mu_{ij} = \log \mu_{i+} + \log \mu_{+j} - \log n. \tag{7.7}$$

Now define the following quantities:

(i) θ to be the average of all the logarithms of cell means, i.e.

$$\theta = \frac{1}{IJ} \sum_{i=1}^{I} \sum_{j=1}^{J} \log \mu_{ij};$$

(ii) α_i to be the difference between the average of the logarithms of means of cells having level i of F_1 and the average of all the logarithms of cell means, i.e.

$$\alpha_i = \frac{1}{J} \sum_{j=1}^{J} \log \mu_{ij} - \theta;$$

(iii) β_j to be the difference between the average of the logarithms of means of cells having level j of F_2 and the average of all the logarithms of cell means, i.e.

$$\beta_j = \frac{1}{I} \sum_{i=1}^{I} \log \mu_{ij} - \theta.$$

Summing (7.7) over both i and j and dividing by IJ gives

$$\theta = \frac{1}{I} \sum_{i=1}^{I} \log \mu_{i+} + \frac{1}{J} \sum_{j=1}^{J} \log \mu_{+j} - \log n. \tag{7.8}$$

Summing (7.7) over j, dividing by J, and subtracting (7.8) gives

$$\alpha_i = \log \mu_{i+} - \frac{1}{I} \sum_{i=1}^{I} \log \mu_{i+}, \tag{7.9}$$

while summing (7.7) over i, dividing by I, and subtracting (7.8) gives

$$\beta_j = \log \mu_{+j} - \frac{1}{J} \sum_{j=1}^{J} \log \mu_{+j}. \tag{7.10}$$

Thus, finally, substituting from (7.9) and (7.10) into (7.7) and using (7.8) gives the model

$$\log \mu_{ij} = \theta + \alpha_i + \beta_j, \tag{7.11}$$

and from (7.9) and (7.10) it follows that $\sum_i \alpha_i = \sum_j \beta_j = 0$.

Expression (7.11) is a *log–linear* model for the cell means μ_{ij} on the hypothesis that F_1 and F_2 are independent. The structure of this model is very similar to the linear models of Chapter 4 for analysis of variance, and the parameters may therefore be given analogous interpretations. We thus refer to θ as the *overall mean*, and to the α_i and β_j as the *main effect terms* of F_1 and F_2 respectively. The α_i reflect the differences in the marginal totals for the rows of the table, while the β_j reflect the differences in the marginal totals for the columns of the table. The constraints on these parameters ensure their estimability.

In general, of course, the two factors need *not* be independent but may exhibit some *association*. In this case the logarithms of the cell means can not be represented simply by an addition of the relevant main effects, as they will be affected differentially by the different combinations of levels of the two factors. In other words, there is an *interaction* between F_1 and F_2, and this interaction must be accommodated within the model. By analogy with analysis of variance again, this suggests that the full log–linear model for cell means in a two-way contingency table should be

$$\log \mu_{ij} = \theta + \alpha_i + \beta_j + \gamma_{ij}, \qquad (7.12)$$

where γ_{ij} represents the interaction term for levels i, j of F_1, F_2 and these terms satisfy the constraints $\sum_i \gamma_{ij} = \sum_j \gamma_{ij} = 0$. Taking all constraints into consideration, this model has $1 + (I - 1) + (J - 1) + (I - 1)(J - 1) = IJ$ unknown parameters. If it is fitted to a $I \times J$ table of frequencies, therefore, a perfect fit is anticipated because the number of items of information (i.e. cell frequencies) exactly matches the number of unknown parameters; this is thus the *saturated* model. To test the null hypothesis of no association between F_1 and F_2, therefore, we need to test whether $\gamma_{ij} = 0$ for all i, j.

Questions of analysis are taken up in the next section. Before getting on to this aspect, however, we consider models for higher-way tables. Essentially, all the above ideas extend readily to higher-way tables, but with inclusion of higher-order interactions as used in standard analysis of variance. Also, some thought is necessary when choosing an appropriate model to fit to the table. Consider, for example, a three-way table classified by factors F_1, F_2 and F_3 having I, J and K levels and indexed by subscripts i, j and k respectively. A saturated log–linear model in this case is given by

$$\log \mu_{ijk} = \theta + \alpha_i + \beta_j + \gamma_k + (\alpha\beta)_{ij} + (\alpha\gamma)_{ik} + (\beta\gamma)_{jk} + (\alpha\beta\gamma)_{ijk} \quad (7.13)$$

where, to avoid proliferation of Greek letters, we denote by $\alpha_i, \beta_j, \gamma_k$ the main effect terms for F_1, F_2, F_3; by $(\alpha\beta)_{ij}, (\alpha\gamma)_{ik}, (\beta\gamma)_{jk}$ the *two-factor interaction* terms for $F_1 \times F_2, F_1 \times F_3$ and $F_2 \times F_3$; and by $(\alpha\beta\gamma)_{ijk}$ the *three-factor interaction* terms for $F_1 \times F_2 \times F_3$. This model carries the constraints

(i) $\sum_i \alpha_i = \sum_j \beta_j = \sum_k \gamma_k = 0.$
(ii) $\sum_i (\alpha\beta)_{ij} = \sum_j (\alpha\beta)_{ij} = \sum_i (\alpha\gamma)_{ik} = \cdots = \sum_k (\beta\gamma)_{jk} = 0.$
(iii) $\sum_i (\alpha\beta\gamma)_{ijk} = \sum_j (\alpha\beta\gamma)_{ijk} = \sum_k (\alpha\beta\gamma)_{ijk} = 0.$

Various situations can then be accommodated by omitting some of the terms from this model, as follows.

1. *Three-factor interaction absent.*
   Under this model, each two-factor interaction is unaffected by the level

of the third factor. There is therefore *partial association* between each pair of factors.

2. *Three-factor and one two-factor interaction absent.*
   There are three versions of this model, depending on which two-factor interaction is absent. If, for example, it is $(\alpha\beta)_{ij}$, then F_1 and F_2 are independent for every level of F_3, but each is associated with F_3 (through the two-factor interactions that are present). Thus F_1 and F_2 are *conditionally independent*, given the level of F_3.

3. *Three-factor and two two-factor interactions absent.*
   Again, there are three versions of this model, this time depending on which two-factor interaction is *present*. If, for example, it is $(\beta\gamma)_{jk}$, then F_2 and F_3 are associated but F_1 is completely independent of both F_2 and F_3.

4. Three-factor and all two-factor interactions absent.
   Here we have *complete independence* between F_1, F_2 and F_3.

Testing for appropriateness of any of these models thus reduces to the testing of zero for the value(s) of the corresponding interaction parameter(s).

Models for tables that have more than three classifying factors are obvious generalizations of the above, and, in particular, the saturated model is obtained in the obvious way from (7.13) by including terms representing main effects and interactions of all orders up to that of the table itself.

7.4 Models for contingency tables

The discussion in Section 7.3 above has established the log–linear predictor as appropriate for the means of the random variables in the cells of the table (i.e. for the systematic component of the overall model), but there is still some uncertainty about the random component. In Section 7.2 we identified three possible sampling schemes that could give rise to an observed table, and found appropriate distributional models for each of these schemes. On the face of it, therefore, it seems as if we need a *set* of models if we are to deal with all eventualities. Fortunately, it has been established by various authors that both the multinomial sampling scheme (when the total sample size is fixed) and the product-multinomial sampling scheme (when specified margins are fixed) are equivalent to the independent Poisson scheme (where nothing is assumed to be fixed) as regards maximum likelihood estimates and test statistics under a log-linear systematic component, provided certain conditions are obeyed. These conditions are that if particular margins are assumed to be fixed, then terms appropriate to those margins are included in the log-linear predictor. For example, if the one-way margin corresponding to factor F_1 is assumed to be fixed, then the terms representing main effects of F_1 *must* be included in the log–linear predictor; if the two-way margin corresponding to $F_1 \times F_2$ is assumed to be fixed, then all terms representing main effects of both F_1 and F_2 as well as terms representing the $F_1 \times F_2$ interaction *must* be included in the log–linear predictor; and so on.

In view of this equivalence between sampling schemes, it is therefore possible to formulate a single model framework for the analysis of contingency tables. We suppose that there are m classifying factors $F_1, \ldots, F_m$ (so that we have an m-way table), but to avoid subscript complexity let us just denote the entries in the table by y_i for $i = 1, \ldots, C$ where C is the total number of cells. In general, the y_i are realizations of random variables Y_i, whose means are μ_i, for $i = 1, \ldots, C$. Moreover, let us follow the practice established in earlier chapters by representing a factor F_i with I levels by $I - 1$ dummy 0/1 variables $Z_1, \ldots, Z_{I-1}$ (see, e.g., Section 3.3.8 for details). A model such as (7.13) can thus be written in the form

$$\log \mu_i = \zeta_1 x_1 + \zeta_2 x_2 + \cdots + \zeta_p x_p,$$

where the ζ_i are either overall mean, or main effect, or interaction terms, and the x_i are the observed values of either 1 (for overall mean), or of the appropriate Z_i (for a main effect term), or of the appropriate product $Z_i Z_j$ (for a two-factor interaction), or of the appropriate product $Z_i Z_j Z_k$ (for a three-factor interaction), and so on.

The following generalized linear model then provides a basis for analysing any contingency table:

(i) The Y_i are mutually independent Poisson random variables with means μ_i for $i = 1, \ldots, C$.

(ii) The classifying factors provide a set of dummy variables $X_1, \ldots, X_p$ which give rise to linear predictors $\eta_i = \beta_1 x_{i1} + \beta_2 x_{i2} + \cdots + \beta_p x_{ip}$ for $i = 1, \ldots, C$; the first variable X_1 is a dummy variable whose values all equal 1 to allow a constant term in the model, while the other variables X_i are either binary 0/1 variables or products of binary 0/1 variables to represent main effects and interactions of the classifying factors.

(iii) The link between (i) and (ii) is that $\log \mu_i = \eta_i$.

There is just one special circumstance that has to be pointed out. Sometimes a table will have cells in which $y_i = 0$. Such zero entries can be of two types: *sampling zeros* or *structural zeros*. A sampling zero is when the frequency of that class is small and we have simply not observed it in the sample; such zeros are perfectly acceptable and should be included in the standard analysis described below. A structural zero, on the other hand, is where that class is impossible so that its actual probability is zero. Such cells must be omitted entirely from the analysis.

Finally, it is worth reiterating the point made in earlier contexts about the use of dummy variables to represent qualitative variables. We are at liberty to choose *any* two distinct values as the indicators, not necessarily just 0 and 1. The only difference will come in the interpretation of the parameters of the model. This point has been illustrated for regression models at the end of Section 3.3.8, and exactly the same considerations apply here for generalized linear models. It should thus be appreciated that different parameterizations can exist for any given situation, depending on what values the dummy variables take and how these values are assigned to the levels of the categorical variables. Correct interpretation of the results requires consideration of these aspects, and we illustrate such considerations in the examples below.

7.5 Analysis

Typical questions of interest that arise when analysing a contingency table, particularly a high-order one, will involve the estimation and testing for significance of main effects and interactions. We wish to know which of these effects are important in explaining the pattern of values in the cells of the table, and to estimate their magnitudes. Also, we are often interested in determining the most parsimonious model that provides a good fit to the data (e.g. whether any of the four special cases listed in Section 7.3 for the three-way table can be used in place of the saturated model 7.13), and this will involve testing whether specified terms in the full model can be set to zero, inspecting various diagnostics such as residuals, and conducting analyses of deviance. However, none of these aspects involves anything fundamentally new, as we have already derived most of the necessary quantities in some of the examples of Chapter 5. We thus collect together the various results in this section.

7.5.1 Estimation and goodness of fit

Relevant features of the generalized linear model specified above have been obtained in Examples 5.4, 5.6 and 5.7. First, we see from Example 5.6 that the log–linear predictor forms a canonical link. Moreover, the same example sets out the iterative scheme for obtaining the maximum likelihood estimates $\hat{\beta}_1, \ldots, \hat{\beta}_p$. Next, Example 5.4 shows that the scale parameter of the model is 1, while Example 5.7 yields the scaled deviance of the model as

$$D = 2 \sum_{i=1}^{C} [y_i \{\log(y_i) - \log(\hat{\mu}_i)\} - y_i + \hat{\mu}_i] = 2 \sum_{i=1}^{C} \left[y_i \log\left(\frac{y_i}{\hat{\mu}_i}\right) - (y_i - \hat{\mu}_i) \right],$$
(7.14)

where $\hat{\mu}_i$ is the set of fitted means

$$\hat{\mu}_i = e^{\hat{\beta}_1 x_{i1} + \ldots + \hat{\beta}_p x_{ip}}.$$

If the table has C cells and the linear predictor has p parameters, then D has an asymptotic χ^2_{C-p} distribution if the model is 'correct'. However, as in previous cases, it should be noted that this distribution may be a poor approximation to the true null distribution of D and a better procedure is via comparison of deviances.

7.5.2 Hypothesis testing and model choice

It has been shown above that certain models of interest arise by setting some of the parameters of the full model to zero. Formal testing of appropriateness of such models can therefore be done in the usual way by analysis of deviance: if the full model has p parameters β_i, and if a submodel of interest is obtained by setting $p - q$ of these β_i to zero (thus leaving q of them to be estimated), then the procedure is to fit the full model and find its scaled deviance D_1, fit the reduced model and find its scaled deviance D_2, and hence obtain the test statistic as the change in deviance $D_2 - D_1$. Under the null hypothesis that the $p - q$ selected β_i are indeed zero, $D_2 - D_1$ has an approximate χ^2_{p-q} distribution, so this can be used to assess the appropriateness of the model.

A sequence of such comparisons of deviance can thus be used to arrive at the most parsimonious model to fit to any given contingency table. When the table is classified by many factors, then reducing the number of parameters to be fitted is very important. As in all other modelling situations considered in previous chapters, this strategy requires us to select *nested* sequences of models for testing, and there may be many possible such sequences in any given situation. There may thus be an element of trial and error in the process of model choice. However, whatever sequence we choose, we should ensure that all fitted models obey the *marginality principle*: whenever a given interaction term is included in a model then all marginal main effect and lower-order interaction terms derivable from it should also be included.

Finally, there is one extra consideration arising from the sampling scheme. We have seen earlier that various margins of the table may be fixed *by design*, and if they are fixed in this way then the terms appropriate to those margins *must* be included in any fitted model (thereby ensuring that observed and fitted values are the same in those margins). Hence if any margins are deemed to be fixed by design, then the *minimal* model that can be fitted is the one that contains parameters appropriate to those margins and any other model has to add parameters to this minimal model.

7.5.3 Model criticism

Having fitted a model, we should inspect aspects of the fit to check on whether it is a reasonable model and whether there are any subareas in which it can be improved. The *fitted values* have been given above already, viz.

$$\hat{\mu}_i = e^{\hat{\beta}_1 x_{i1} + \cdots + \hat{\beta}_p x_{ip}},$$

and from these we can derive various types of *residuals*. The two most useful types have been defined in general terms in Chapter 5, the *standardized Pearson residual* by

$$r_{i(P)} = \frac{y_i - \hat{\mu}_i}{\sqrt{[(1 - h_i)\widehat{\text{var}(y_i)}]}} \tag{7.15}$$

and the *standardized deviance residual* by

$$r_{i(D)} = \frac{\text{sgn}(y_i - \hat{\mu}_i)\sqrt{D_i}}{\sqrt{(1 - h_i)}} \qquad (7.16)$$

where D_i is the contribution made to the scaled deviance by observation i and $\text{sgn}(y_i - \hat{\mu}_i)$ is the sign of $(y_i - \hat{\mu}_i)$. Turning once again to Example 5.4, we see that in the present case we have $\widehat{\text{var}}(y_i) = \hat{\mu}_i$ for substitution in (7.15). Moreover, although the standardized versions of these residuals are the best ones to use, the presence of the term $1 - h_i$ in each means that the standardized residuals have to be individually evaluated for each new fitted model (since h_i depends on the $\hat{\beta}_i$ present in the model as well as on the design matrix X). It is thus computationally convenient to drop this term, and to calculate instead either the *scaled Pearson residual*

$$r'_{i(P)} = \frac{y_i - \hat{\mu}_i}{\sqrt{\hat{\mu}_i}} \qquad (7.17)$$

or the *scaled deviance residual*

$$r'_{i(D)} = \text{sgn}(y_i - \hat{\mu}_i)\sqrt{D_i}. \qquad (7.18)$$

This simplification introduces a degree of approximation to the estimated variances of the residuals, but results are generally sufficiently accurate for any drawback to be outweighed by the computational benefits.

We thus have all the necessary mechanisms at our disposal for analysing contingency tables of any size and shape. The following examples illustrate various situations and problems that typically arise in practice.

7.6 Applications

Example 7.1. We start with a simple two-way table. The values in Table 7.4 give incidences of the various ABO red cell blood groups in samples from five different racial/language groups in the Middle East. An obvious question to ask here is whether the the blood group frequencies can be assumed to be the same in each population from which the samples have been taken. If they can be so assumed, then we say that there is *no association* between the two classifying factors (race/language and blood group).

Turning to the model, we recollect that the saturated log–linear predictor for a two-way table is given by (7.12). Absence of association between the factors implies that there should be no interaction terms between the two factors in this predictor, so to answer the question we need to establish whether predictor (7.11) fits the data as well as does (7.12). However, to express these predictors in the manner of Section 7.4, we need to define the relevant dummy variables Z_i and parameters β_i. This is done as follows.

Table 7.4 *Samples from five racial/language groups, classified according to their ABO red cell blood groups.*

Race/language	O	A	B	AB	Totals
		Blood Group			
Kurd	531	450	293	226	1500
Arab	174	150	133	36	493
Jew	42	26	26	8	102
Turkoman	47	49	22	10	128
Osmanli	50	59	26	15	150
Totals	844	734	500	295	2373

(i) Overall mean. This just requires a single dummy variable which takes value 1 in all cells, so we can take $X_1 = Z_1$ with $x_1 = 1$ for all cells.

(ii) Blood group. Here we have 4 levels (O, A, B and AB) so we need three dummy variables Z_2, Z_3, Z_4. Various assignments of values are possible, of course, but for consistency with all previous assignments we set $Z_2 = Z_3 = Z_4 = 0$ in cells corresponding to the first level O, $Z_2 = 1, Z_3 = Z_4 = 0$ in cells corresponding to the second level A, $Z_2 = 0, Z_3 = 1, Z_4 = 0$ in cells corresponding to the third level B, and $Z_2 = Z_3 = 0, Z_4 = 1$ in cells corresponding to the fourth level AB.

(iii) Race/language. Here there are five levels, so we need four dummy variables Z_5, Z_6, Z_7, Z_8. By analogy with the above, we set $Z_5 = Z_6 = Z_7 = Z_8 = 0$ for cells involving Kurds, $Z_5 = 1, Z_6 = Z_7 = Z_8 = 0$ for cells involving Arabs, $Z_5 = 0, Z_6 = 1, Z_7 = Z_8 = 0$ for cells involving Jews, and so on.

Having defined the Z_i, we can now specify the terms of the model. The overall mean is given by the term $\beta_1 x_1$. The blood group main effect has three degrees of freedom and is given by $\beta_2 x_2 + \beta_3 x_3 + \beta_4 x_4$ where $X_2 = Z_2$, $X_3 = Z_3$, $X_4 = Z_4$, while the race/language main effect has four degrees of freedom and is given by $\beta_5 x_5 + \cdots + \beta_8 x_8$ where $X_5 = Z_5, \ldots, X_8 = Z_8$. If we then wish to go on and include interaction terms, we need to include products of the Z_i. For example, we could add terms $\beta_9 x_9$ with $X_9 = Z_2 Z_5$, $\beta_{10} x_{10}$ where $x_{10} = Z_2 Z_6$, and so on. The saturated model would contain all 12 terms that consist of each 'blood group' Z_i multiplied by each 'race/language' Z_i in this way.

We note that by defining the x_i as above, the lowest level of each factor provides the 'baseline'. Thus the cell corresponding to blood group O for Kurds has $\log \mu_i = \beta_1$, the overall mean. Then the cell corresponding to blood group A for Kurds has $\log \mu_i = \beta_1 + \beta_2$ and so on. Each β_i therefore indicates the effect on the logarithm of the cell mean of that level *relative to the lowest level* for the relevant factor. This parameterization ensures the correct number of β_i for estimability, so no further constraints are necessary.

Having specified the necessary dummy variables, the 'no association' log–

linear predictor is given by

$$\log \mu_i = \beta_1 x_1 + \beta_2 x_2 + \cdots + \beta_8 x_8,$$

and when we fit this model we find the following parameter estimates and standard errors:

Parameter	Estimate	Standard Error
β_1	6.279	0.0378
β_2	−0.1396	0.0505
β_3	−0.5235	0.0564
β_4	−1.051	0.0676
β_5	−1.113	0.0519
β_6	−2.688	0.1023
β_7	−2.461	0.0921
β_8	−2.303	0.0856

The fitted values $\hat{\mu}_i$ defined in (7.14) are given in the second line in each cell of Table 7.5, and hence the scaled deviance for this model can be shown from (7.14) to be equal to 45.414. This has 12 degrees of freedom (the number of interaction terms specified above that have been omitted from the saturated model), so is highly significant. The conclusion is therefore that some or all of the interaction terms are *not* equal to zero, and hence that there *is* evidence of association between the two factors.

Note that in a two-way table, the simple chi-squared test of association can also be applied to answer the same question. Here we obtain a value $X^2 = 44.43$ of the test statistic, also on 12 degrees of freedom. This statistic has a chi-squared distribution under the null hypothesis of no association. The present value is thus clearly significant, which means that we reject the null hypothesis and reach the same conclusion as we did using the deviance.

In order to examine where in the table the simple additive model breaks down and where there is evidence of interaction, we need to calculate either the scaled Pearson residuals or the scaled deviance residuals. These have both been calculated for the blood group data and are shown in lines 3 and 4 of each cell of Table 7.5; for completeness the observed values are reproduced in the top line of each cell of this table. It is evident from these values that a major discrepancy occurs in blood groups B and AB for Kurds and Arabs relative to the rest of the values, since values of around 3 or more indicate large scaled residuals. A simple additive structure on the log–linear scale is therefore not an adequate model for these data, and some interaction terms need to be included if the fit is to be a good one.

Example 7.2. Next we consider a simple example of a three-way table, namely the data given in Table 7.2 on survival of babies. The three factors are 'clinic', 'care' and 'survival' and each has two levels thus yielding a $2 \times 2 \times 2$ table. We

Table 7.5 *Observed values y_i, fitted values $\hat{\mu}_i$, scaled Pearson residuals $r'_{i(P)}$ and scaled deviance residuals $r'_{i(D)}$ for the blood group data.*

Race/language		Blood Group			
		O	A	B	AB
Kurd	y_i	531	450	293	226
	$\hat{\mu}_i$	533.50	463.97	316.06	186.47
	$r'_{i(P)}$	−0.1083	−0.6485	−1.2969	2.8946
	$r'_{i(D)}$	−0.1083	−0.6519	−1.3132	2.8004
Arab	y_i	174	150	133	36
	$\hat{\mu}_i$	175.34	152.49	103.88	61.29
	$r'_{i(P)}$	−0.1015	−0.2017	2.8574	−3.2301
	$r'_{i(D)}$	−0.1016	−0.2023	2.7375	−3.5024
Jew	y_i	42	26	26	8
	$\hat{\mu}_i$	36.28	31.55	21.49	12.68
	$r'_{i(P)}$	0.9500	−0.9881	0.9725	−1.3143
	$r'_{i(D)}$	0.9265	−1.0194	0.9411	−1.4109
Turkoman	y_i	47	49	22	10
	$\hat{\mu}_i$	45.53	39.59	26.97	15.91
	$r'_{i(P)}$	0.2185	1.4952	−0.9570	−1.4822
	$r'_{i(D)}$	0.2174	1.4411	−0.9889	−1.5920
Osmanli	y_i	50	59	26	15
	$\hat{\mu}_i$	53.35	46.40	31.61	18.65
	$r'_{i(P)}$	−0.4587	1.8502	−0.9971	−0.8446
	$r'_{i(D)}$	−0.4636	1.7747	−1.0290	−0.8747

will again define the dummy variable x_1 with all values equal to 1 to correspond to the overall mean, but now we need just a single binary dummy variable to represent the two levels of each factor. We define x_2 to represent 'clinic' (value 0 for clinic A, value 1 for clinic B), x_3 to represent 'care' (value 0 for low level, value 1 for high level), and x_4 to represent 'survival' (value 0 for survived, value 1 for died).

First consider testing for association in each clinic separately. This requires us to fit the model with $\log \mu_i = \beta_1 x_1 + \beta_3 x_3 + \beta_4 x_4$ and Poisson errors to each 2×2 subtable as in Example 7.1 above, and we find scaled deviances of 0.0822 and 0.000096 respectively. Each of these has one degree of freedom and each is very evidently not significant, so we conclude that there is no association between care and survival in each clinic separately. We might thus consider pooling the data in the two clinics to obtain the *collapsed* table:

Care	Survival Survived	Died	Totals
Low	373	20	393
High	316	6	322
Totals	689	26	715

However, fitting the 'no association' model to this collapsed table yields a deviance of 5.612 on 1 degree of freedom which is significant. This model thus does not fit adequately, suggesting that survival now depends on level of care. This apparent paradox shows the danger of collapsing tables over factors; we will return to this point at the end of the example.

For the present, it seems as if we cannot readily simplify the table so we must analyse the full $2 \times 2 \times 2$ arrangement. Looking at the full table, and calculating survival percentages, the patterns are clear: little effect of level of care, but a large difference between clinics. Let us therefore consider a logical hierarchy of models, starting with the simplest, until we find one that fits adequately.

The simplest model assumes complete independence between factors, i.e.

$$\log \mu_i = \beta_1 x_1 + \beta_2 x_2 + \beta_3 x_3 + \beta_4 x_4.$$

This model yields a very large scaled deviance of 211.48 on 4 degrees of freedom, indicating that it does not fit. This result was to be expected, of course, from the preliminary calculations above. However, since we have a term for each main effect in the model, we might note that all the fitted one-way margins will agree with their observed counterparts.

Next, our preliminary calculations above have suggested that survival seems to depend on clinic, so we should add a term representing the 'survival $\times$ clinic' interaction to the model. Since the factors are all at two levels this interaction has 1 degree of freedom and is represented by the single multiplicative term $\beta_{24} x_2 x_4$, so we fit the model

$$\log \mu_i = \beta_1 x_1 + \beta_2 x_2 + \beta_3 x_3 + \beta_4 x_4 + \beta_{24} x_2 x_4.$$

The scaled deviance now is 193.74 on 3 degrees of freedom. This is a large (significant) reduction on the previous deviance ($211.48 - 193.74 = 17.74$ on 1 degree of freedom), but still a very high absolute value which indicates a poor fit to the data. We therefore need to add further terms to the model. An obvious candidate is the 'care $\times$ clinic' interaction, as there are some evident differences in care levels between clinics. This interaction is represented by the multiplicative term $\beta_{23} x_2 x_3$, so we fit the model

$$\log \mu_i = \beta_1 x_1 + \beta_2 x_2 + \beta_3 x_3 + \beta_4 x_4 + \beta_{23} x_2 x_3 + \beta_{24} x_2 x_4.$$

The scaled deviance for this model is 0.0823 on 2 degrees of freedom, not only a huge decrease from the previous value but also a very satisfactory (non-significant) absolute value indicating that the model fits well. This therefore seems to be the best model.

However, there is just one more aspect to check. Perhaps with the inclusion of the 'care × clinic' interaction we no longer need to keep the 'survival × clinic' interaction previously included. If it can be dropped without detriment, then we will have a more parsimonious model. To check on this, we fit the model

$$\log \mu_i = \beta_1 x_1 + \beta_2 x_2 + \beta_3 x_3 + \beta_4 x_4 + \beta_{23} x_2 x_3$$

and obtain a scaled deviance of 17.828 on 3 degrees of freedom. This is now an increase of 17.75 on 1 degree of freedom as well as being a high deviance value in its own right. The conclusion is thus that we *cannot* drop the 'survival × clinic' interaction, so the best model for the data is indeed

$$\log \mu_i = \beta_1 x_1 + \beta_2 x_2 + \beta_3 x_3 + \beta_4 x_4 + \beta_{23} x_2 x_3 + \beta_{24} x_2 x_4.$$

Parameter estimates for this model are

Parameter	Estimate	Standard Error
β_1	0.9679	0.38258
β_2	1.866	0.4466
β_3	0.5063	0.0946
β_4	4.205	0.3807
β_{23}	−2.653	0.2316
β_{24}	-1.756	0.4496

and the table of observed (y_i) and fitted ($\hat{\mu}_i$) values is

Clinic	Antenatal Care		Survived	Died	Totals
A	Low	y_i	176	3	179
		$\hat{\mu}_i$	176.368	2.632	179
	High	y_i	293	4	297
		$\hat{\mu}_i$	292.632	4.368	297
B	Low	y_i	197	17	214
		$\hat{\mu}_i$	196.987	17.013	214
	High	y_i	23	2	25
		$\hat{\mu}_i$	23.013	1.987	25
	Totals	y_i	689	26	715
		$\hat{\mu}_i$	689	26	715

It is evident from these fitted values that the model fit is very good, so we do not need to display any residuals. Note also that the observed and fitted values agree in the two-way margins for 'care × clinic' and 'survival × clinic' as well as for each of the three one-way margins, reflecting the terms included in the final model.

Finally, we return to the question of collapsability of the table over factors or combinations of factors. The reason why we could not collapse the table over

Table 7.6 *Incidence of* Torus Mandibularis *in three Eskimo populations; presence denoted by '+', absence by '−' (Source: Muller and Mayhall, 1970).*

Popn	Sex		1–10	11–20	21–30	31–40	41–50	51+	Tot
Iglook	M	+	4	8	13	18	10	12	65
		−	44	32	21	5	0	1	103
	F	+	1	11	19	13	6	10	60
		−	42	17	17	5	4	2	87
Beech	M	+	2	5	7	5	4	4	27
		−	17	10	6	2	2	1	38
	F	+	1	3	2	5	4	2	17
		−	12	16	6	2	0	0	36
Aleut	M	+	4	2	4	7	4	3	24
		−	6	13	3	3	5	3	33
	F	+	3	1	2	2	2	4	14
		−	10	7	12	5	2	1	37
Totals			146	125	112	72	43	43	541

the factor 'clinic' (i.e. pool the data for the two clinics) is that 'clinic' interacts with *each* of the two other factors (as evidenced by the need to include both interaction terms in the model). The general rule is that we can collapse over a factor as long as that factor is independent of *at least one* of the other factors in the study, but care should always be taken when trying to simplify tables in this way.

Example 7.3. This next example illustrates a systematic approach to model choice when we have a large table for which there are so many models that could be fitted that it is impracticable to fit them all and choose the best. Table 7.6 shows the incidence of the morphological trait *Torus Mandibularis* (a small protuberance in the lower jaw at the front of the mouth) in three Eskimo populations, classified also by age and sex to yield a $3 \times 2 \times 2 \times 6$ table.

We see that there are 6 two-factor interactions, and hence a possible 64 models even before we consider any three-factor interactions, so we need some systematic approach to the selection of terms for the model.

One important distinction we should make, which will help in model choice, is between response and explanatory variables. We are generally interested in finding any interactions between response and explanatory variables; interactions among the explanatory variables themselves are in a sense 'nuisance' effects which, although they cannot be removed, are not in themselves of interest. In Table 7.6 the response variable is incidence (i.e. presence/absence of *Torus Mandibularis*), while the other factors are all explanatory.

A second important consideration in general is whether the sampling design requires certain margins to be fixed, because this will determine those terms that *must* be included in any fitted model. Let us assume in the present example that a fixed sample size was taken from each population but no other values were fixed in advance. This means that the factor 'population' must appear in any fitted model (thereby ensuring that any fitted values produce the one-way margin of population totals with values 315, 118, 108), but otherwise a free choice of terms is allowed.

Apart from these considerations, we proceed by fitting what seems to be the most reasonable hierarchical sequence of models, testing additional terms for significance each time and if necessary inspecting tables of residuals to give suggestions for further terms to include, until we obtain a model that fits the data well. We can then check to see if any terms can be omitted from this model, in order to achieve the greatest parsimony.

Let us therefore turn to the Eskimo data. For simplicity we will forgo the detailed structure in terms of dummy variables, but simply denote the three explanatory factors by POP (for population), SEX and AGE, and the response variable by INC (for incidence). In the absence of further information, the simplest place to start is with the model of complete independence. Assuming that the overall mean is always included, this model can be denoted by

$$POP + SEX + AGE + INC$$

and it yields a scaled deviance of 225.74 on 62 degrees of freedom; clearly a poorly fitting model.

Since there are now many possible models that could be tried in a 'forward selection' fashion, a more fruitful approach might be to go to the 'largest' non-saturated model and work backwards instead. The model that includes all terms apart from the four-factor interaction $POP.SEX.AGE.INC$ has scaled deviance 21.607 on 10 degrees of freedom. This is significant at about the 2% level, so is not very good. We first try to simplify it. Since the three-factor interactions involving both SEX and INC are of special interest, we look to see if we can remove any of them. Eliminating $POP.SEX.INC$ increases the scaled deviance by only 1.995 on 2 degrees of freedom, and then eliminating $AGE.SEX.INC$ further increases it by 4.134 on 5 degrees of freedom, so neither of these interactions is significant and both can be removed. Moreover, eliminating the two-factor interaction $SEX.INC$ then changes the deviance by only 0.959 on 1 degree of freedom, so this interaction can also be removed. At this stage we have the model

$$POP + AGE + SEX + INC + POP.AGE + POP.SEX + AGE.SEX$$
$$+POP.INC + AGE.INC + POP.AGE.SEX + POP.AGE.INC$$

Table 7.7 *Scaled Pearson residuals for the model in the text and the data of Table 7.6.*

Popn	Sex		1–10	11–20	21–30	31–40	41–50	> 50
					Age Groups			
Iglook	M	+	0.84	−0.95	−0.65	0.15	0.71	0.17
		−	−0.20	0.59	0.59	−0.26	−1.41	−0.45
	F	+	−0.88	1.14	0.63	−0.17	−0.71	−0.17
		−	0.21	−0.71	−0.58	0.29	1.41	0.47
Beech	M	+	0.16	0.78	0.61	0.00	−0.37	−0.14
		−	−0.05	−0.43	−0.52	0.00	0.73	0.34
	F	+	−0.20	−0.70	−0.77	0.0	0.45	0.22
		−	0.06	0.39	0.67	0.0	−0.89	−0.53
Aleut	M	+	0.55	0.03	1.41	0.74	−0.07	−0.42
		−	−0.36	−0.01	−0.89	−0.79	0.07	0.55
	F	+	−0.48	−0.04	−1.00	−0.88	0.11	0.46
		−	0.31	0.02	0.63	0.94	−0.10	−0.61

and this has scaled deviance 28.694 on 18 degrees of freedom. This value is not significant at the 5% level, so we seem to have obtained the best model (noting also that it satisfies the marginality principle). Just to check, we calculate the scaled Pearson residuals and show them in Table 7.7. It is evident that there are no particularly large values or obvious patterns in this set of residuals, so we can conclude that the model does fit the data. Moreover, omitting any of the terms remaining in the model leads either to a significant scaled deviance or to obvious patterns of large residuals, so the above model is in fact the best one for the data.

As a final comment, we note that the analysis here has followed the structure developed in the present chapter for *counts* and has therefore involved the fitting of a log-linear model to a $3 \times 2 \times 2 \times 6$ set of independent Poisson variables. However, the data could equally well be viewed as a set of *proportions* of Eskimos exhibiting *Torus Mandibularis* in each cell of a $3 \times 2 \times 6$ table (classified by population, sex and age group). If this view is taken, then the data could be analysed by the methods of Chapter 6, i.e. by fitting logistic regressions to these proportions (using the same linear predictors but assuming binomial errors). In fact, the two analyses are exactly equivalent and it is up to the analyst to choose whichever is the more convenient (either from computational or interpretational perspectives).

Example 7.4. This last example briefly illustrates how we can handle factors that have numeric values associated with their levels. Consider the data in Table 7.3 on coal miners classified by age, breathlessness and wheeze. There are 9 age groups, and each of breathlessness and wheeze is a 'presence/absence' binary factor, so we have a $9 \times 2 \times 2$ table.

We fit the model with all possible terms except the three-factor $AGE \times BREATH \times WHEEZE$ interaction, i.e.

$$\log \mu_{ijk} = \theta + A_i + B_j + W_k + (AB)_{ij} + (AW)_{ik} + (BW)_{jk},$$

where we replace α, β, γ in the notation of equation (7.13) by the first letter of each factor. This model gives a scaled deviance of 26.69 on 8 degrees of freedom, which suggests that a three-factor interaction *is* present, i.e. that the association of wheeze and breathlessness changes between age groups.

Including the full three-factor interaction on 9 degrees of freedom produces the saturated model, however, so is non-informative. On the other hand, the factor AGE is actually quantitative with equal intervals for each of its nine levels. These intervals can therefore be coded $1, 2, \ldots, 9$, and we might look to see if the $BREATH \times WHEEZE$ interaction changes linearly with age. To do this we simply include a term $(BW)_{jk}x_i$, where x_i takes one of the values $1, \ldots, 9$ according to the age group of the cell in question. This is just a single extra term (since wheeze and breathlessness each just have two levels), and so represents 1 of the available 9 degrees of freedom in the full three-factor interaction. Fitting this model gives a scaled deviance of 6.8 on 7 degrees of freedom, a perfectly acceptable fit to the data. The three-factor interaction can thus be explained by this simple linear structure with age.

8
Further topics

8.1 Introduction

In preceding chapters we have described various linear and generalized linear
models that can be fitted to data, and have developed techniques of analysis
appropriate to these models. The range of models and techniques has been
such as to cover most of the common types of problem that arise in practice.
However, this range is by no means exhaustive, and many problems do arise
that require more advanced methodology than can be covered in an introductory
text such as this one. In this final chapter, therefore, we gather together a
number of further topics for brief mention. We can do no more than introduce
the main aspects of each topic, point the way in which analysis can proceed, and
refer the reader to more specialized or more advanced texts for further details.
The concluding section then provides a brief discussion of the whole process of
model building in data analysis.

8.2 Continuous non-normal responses

The classical linear model theory outlined in Chapters 3 and 4 for continuous
responses assumes normality and homogeneity of variance, and it was shown
in Section 5.6 that the same results can be achieved by fitting generalized linear models with identity link and normal errors to the data. However, many
situations arise in practice where the response variable is continuous but where
normality or variance homogeneity assumptions are questionable.

In general, there are many reasons why a particular set of data might not
be assumed to arise from a single homogeneous normal distribution, and individual situations often need individual modelling applied to them. However,
there is one class of non-normal continuous responses that is sufficiently widely
encountered for a brief consideration to be worthwhile here. This is the class
in which

- the response variable Y takes only positive values,
- the probability that $Y \geq k$ steadily decreases as k increases,

- the variance of Y is approximately proportional to the square of its mean (i.e. the coefficient of variation of Y is approximately constant).

Examples where such a response might be observed include amounts of rainfall in a series of rainy days, lifetimes of electronic components, and yields of plants in certain situations in agriculture.

An appropriate distribution that has plenty of flexibility and exhibits the right sort of characteristics in such situations is the *gamma* distribution, which has been defined in Section 2.3.2. This is a two-parameter distribution, and we can choose a variety of forms depending on the parameter definitions. If we choose its density in terms of parameters μ, ϕ to be

$$f(y) = \frac{1}{\Gamma(\phi)} \left(\frac{\phi}{\mu} \right)^{\phi} y^{\phi-1} e^{-\phi y/\mu} \qquad (0 \le y < \infty), \qquad (8.1)$$

then we see that

$$f(y) = \exp\left\{ \left(-\frac{y}{\mu} - \log\mu \right)\phi - \log\Gamma(\phi) + \phi\log\phi + (\phi - 1)\log y \right\}. \quad (8.2)$$

This is of the same form as equation (5.1) for $\theta = -1/\mu$, $b(\theta) = -\log(-\theta)$ and $a(\phi) = 1/\phi$. It thus follows from (5.1), (5.2) and (5.3) that Y is a member of the exponential family with canonical parameter θ, mean $E(Y) = b'(\theta) = \mu$, variance function $b''(\theta) = \mu^2$ and hence variance $\text{Var}(Y) = a(\phi)b''(\theta) = \mu^2/\phi$.

Now let us suppose that $Y_1, Y_2, \ldots, Y_n$ are independent responses with distributions of the form of (8.2) but with individual parameters $\mu_1, \mu_2, \ldots, \mu_n$ respectively and a common second parameter ϕ. Moreover, let $X_1, X_2, \ldots, X_p$ be explanatory variables with values $x_{i1}, x_{i2}, \ldots, x_{ip}$ corresponding to Y_i such that μ_i is determined by the linear predictor $\eta_i = \beta_1 x_{i1} + \cdots + \beta_p x_{ip}$ through the link function $g(\mu_i) = \eta_i$. Then, *if ϕ is a known constant* we have all the ingredients for a generalized linear model with gamma variables. From above we see that the canonical link is the inverse function $\eta_i = \mu_i^{-1}$, i.e.

$$\mu_i = \frac{1}{\beta_1 x_{i1} + \cdots + \beta_p x_{ip}},$$

but other commonly used link functions are the log link $\eta_i = \log\mu_i$, i.e.

$$\mu_i = \exp(\beta_1 x_{i1} + \cdots + \beta_p x_{ip}),$$

or the identity link

$$\mu_i = \beta_1 x_{i1} + \cdots + \beta_p x_{ip}.$$

If ϕ is a known constant, then the log-likelihood is

$$l(\mu_1, \ldots, \mu_n) = \phi \sum_{i=1}^{n} (-y_i/\mu_i - \log\mu_i) + C.$$

The maximum of the log-likelihood occurs for the saturated model $\tilde{\mu}_i = y_i$ for all i, and has value $-\phi \sum(1 + \log y_i)$ (provided that we exclude the possibility that $y_i = 0$ for any i). Thus, if all observations are strictly positive, then the (scaled) deviance of any fitted model is

$$D = -2\phi \sum_{i=1}^{n} \{\log(y_i/\hat{\mu}_i) - (y_i - \hat{\mu}_i)/\hat{\mu}_i\} \tag{8.3}$$

where $\hat{\mu}_i = g^{-1}(\hat{\beta}_1 x_{i1} + \cdots + \hat{\beta}_p x_{ip})$ and the parameter estimates $\hat{\beta}_i$ are obtained via the iterative scheme of Section 5.3. Furthermore, from the theory of the same section it can be shown that the approximate covariance matrix of the estimates $\hat{\beta}_i$ is given by $(\phi X' W X)^{-1}$, where X and W are the design and weights matrices (see equation 5.9) evaluated at $\hat{\beta}$. If ϕ is known then $\text{cov}(\hat{\beta})$ can be evaluated exactly, but if ϕ is unknown it must be estimated before $\text{cov}(\hat{\beta})$ can be found. Problems arise with maximum likelihood estimation in this case, but it can be shown that the simple goodness-of-fit statistic $\frac{1}{n-p} \sum\{(y_i - \hat{\mu}_i)/\hat{\mu}_i\}^2$ provides a consistent estimator of ϕ^{-1}.

Further details of this model, and several examples of its application, are given in Chapter 8 of the monograph *Generalized Linear Models* (2nd edition, Chapman & Hall, 1989) by P. McCullagh and J.A. Nelder. Unfortunately, the model does have some severe drawbacks as regards practical applications. One is the already evident problem with the parameter ϕ, which is assumed to be known for the analysis derived from a generalized linear model framework; this analysis is complicated by the presence of unknown ϕ. However, more problematic is the fact that the observations $y_1, \ldots, y_n$ are rarely obtained as cleanly in practice as is suggested above. The most common studies giving rise to observations of the form described above are either those concerned with *survival* data or those concerned with *reliability* data. The former are mostly prevalent in medical applications, where interest focusses on the time elapsing between one event (e.g. administration of some treatment to a patient) and another (e.g. re-emergence of symptoms, or death). The latter occur mainly in technological or engineering applications, where interest focusses on the time elapsing before some individual component either fails or exhibits some symptom of malfunction.

The main practical problem with such studies is that they must necessarily take only a finite, usually pre-specified, time and it cannot be guaranteed that all observations will be completed within this time. Thus, for example, a study of the effect of some new treatment for a particular type of cancer might be scheduled to last a specified number of years. Patients enter the trial at various time points during this period, and the response variable Y is survival time to death of the patient. Some patients will die during the period of the trial, and for them the survival time y_i is well defined. However, others may enter, be treated, and still be alive at the end of the trial. The measured survival time y_i for such individuals, from treatment to end of trial, has therefore been *(right)-censored* and the true survival time is some unknown value $Y \geq y_i$. Similar characteristics occur in reliability studies: some components will malfunction within the designated observation period, and their values y_i represent actual lifetimes, but others are still working correctly at the end of the observation period so that their values y_i are also right-censored and the true lifetimes are unknown values exceeding the observed y_i. *Left-censoring*, though less common, will arise in studies where a designated time must elapse before recordings are taken, so any individual not surviving this initial period is left-censored.

A full analysis of such studies must be able to use the information from the censored observations as well as from the complete ones; indeed, in many studies the censored observations may be in the majority and they actually carry considerable information. Unfortunately, the generalized linear model with gamma errors cannot cope with such censoring and more complicated models need to be introduced.

One aspect of any such model that needs specification is the distribution from which the observations have been assumed to come, and the exponential, the Weibull, and the extreme-value distributions are three commonly-used distributions in survival and reliability studies. However, rather than focussing on either the density function $f(y)$ or the distribution function $F(y)$ (see Section 2.1) of the chosen probability model, in survival and reliability studies we are more concerned with the *hazard* function $h(y)$ defined by

$$h(y) = f(y)/\{1 - F(y)\}.$$

Simple probability theory shows that $h(y)$ measures the probability that an individual does not survive the next small time interval given that the individual has so far survived for a time y.

The second main component of a model for survival/lifetime data is the nature of dependence between the hazard function and the covariates $X_1, \ldots, X_p$ that are associated with the response variable Y for each individual. Here we have another complicating feature, as survival data are frequently associated with *time-dependent* covariates, and we have to allow for individual covariate values x_{ij} changing over time. A common model that is adopted in medical studies is the *proportional hazards* model, in which the hazard function is expressed as the product of two factors one of which depends only on the time y at which the observation is made and the other depends on a linear predictor of the covariates. In reliability studies, an alternative model that is equally popular is the *accelerated failure time* model, in which explanatory variables are assumed to act multiplicatively on the time scale.

The upshot of these complicating features is that models for survival data or for reliability data need to be considerably more complex than the other models considered in this book. Such models involve a number of extra technical aspects, and space restrictions preclude their consideration here. The interested reader is referred for further details to the monographs by D. Collett (*Modelling Survival Data in Medical Research*, Chapman & Hall, 1994) and Crowder, M.J.,

Kimber, A.C., Smith, R.L. and Sweeting, T.J. (*Statistical Analysis of Reliability Data*, Chapman & Hall, 1991).

8.3 Quasi-likelihood

All the methodology associated with the generalized linear model, and the consequent techniques described in the preceding three chapters, have rested on assuming that the response variable Y has a distribution belonging to the exponential family. Such a distribution has a density of the form in equation (5.1), and if $Y_1, Y_2, \ldots, Y_n$ are independent response vectors with densities of this form then the likelihood is given by equation (5.4). This equation is the linch-pin of the methodology, because it leads to the maximum likelihood equations (5.8) for the unknown parameters in the linear predictor of the model, to the likelihood ratio test statistic and hence deviance of equation (5.11) for testing the fit of the model, and to the testing of equivalence of sub-models using the difference of deviances as in Section 5.5.

In particular, let us focus for a moment on the estimation aspects. Recalling the theoretical framework of Chapter 5, suppose we have a set of n random variables $Y_1, Y_2, \ldots, Y_n$ that satisfy the conditions (i) to (iii) of a generalized linear model as given in section 5.2.2. The parameter θ depends on the explanatory variables $X_1, X_2, \ldots, X_p$ so can take different values for different Ys. Also, we assume ϕ to be constant across the Ys, but permit the function $a(\cdot)$ to vary. Thus we write θ_i and $a_i(\phi)$ for these quantities when applied to Y_i ($i = 1, \ldots, n$). From (5.1) it then follows that the likelihood is given by

$$L(\theta_1, \ldots, \theta_n; \phi) = \prod_{i=1}^{n} \exp\{[y_i\theta_i - b(\theta_i)]/a_i(\phi) + c(y_i, \phi)\}, \qquad (8.4)$$

and the theory of Section 5.3 leads to maximum likelihood estimators of the parameters β_i of the linear predictor given by those values which satisfy the equations

$$\sum_{i=1}^{n} \frac{(y_i - \mu_i)x_{ij}}{a_i(\phi)V(\mu_i)g'(\mu_i)} = 0 \qquad \text{for } j = 1, 2, \ldots, p. \qquad (8.5)$$

Sometimes we may not know (or be able to specify as an assumption) the form of density function of the response variable Y, but we may nevertheless have some partial information on this variable by way of formulae for its mean and variance. This information may have come from previous practical experience of the situation, or from theoretical considerations which stop short of specification of the exact probability model. On the other hand, the left-hand sides of the equations in (8.5) depend only on the observed values of the response (y_i) and explanatory (x_{ij}) variables, on their means μ_i and variances $a_i(\phi)V(\mu_i)$, and on the link function $g(\mu_i)$ of the model. Even in the absence of any distributional assumptions, therefore, we can still formulate these *estimating equations* and solve them to obtain estimates $\hat{\beta}_i$ of the linear predictor. Such estimates are known as *quasi-likelihood* estimates, to emphasize that they have not come from a full likelihood analysis.

Estimation proceeds iteratively as before, via either the Newton–Raphson or the scoring methods (see Section 5.3), and we can derive asymptotic covariance matrices for the estimated parameters, goodness-of-fit statistics (called quasi-deviances by analogy with the above terminology), and tests of adequacy of sub-models, in exactly parallel fashion to that for the complete likelihood methods; see Chapter 9 of *Generalized Linear Models* by P. McCullagh and J.A. Nelder (1989, Chapman & Hall) for full technical details. One of the main uses of quasi-likelihood is in the handling of overdispersion (see below). Another example of its use is in the treatment of continuous response variables which take the form of proportions lying between 0 and 1: quasi-likelihood analysis shows that such situations can be handled by the logistic regression methods of Chapter 6, simply replacing the previous binomial parameters π_i and n_i by the observed proportions and sample sizes respectively.

8.4 Overdispersion

We saw in Section 2.3.2 that the normal distribution is completely determined by its two parameters, the mean μ and the variance σ^2. All models for normally-distributed response variables, such as the regression models of Chapter 3 or the analysis of variance models of Chapter 4, therefore include a separate variance parameter. This variance parameter can then be estimated from the data in addition to any linear predictor for the mean, and any degree of variability about the mean is thereby accommodated in the fitted model.

Non-normal distributions, by contrast, do not necessarily have such an independent variance parameter, and those that depend on only one parameter will typically exhibit a fixed relationship between mean and variance. For example, the binomial distribution with parameters n and π has mean $\mu = n\pi$ and variance $\sigma^2 = n\pi(1 - \pi) = \mu(1 - \mu/n)$, while the Poisson distribution with parameter λ has $\mu = \sigma^2 = \lambda$. In each of these cases, therefore, determining the mean of the distribution implicitly determines its variance also and there is no flexibility in the matter.

This feature has implications when fitting generalized linear models based on either the binomial or the Poisson distribution, such as those described in Chapters 6 and 7. Sometimes a particular distribution may seem to be eminently suitable for a response variable from either theoretical or empirical considerations, but when we inspect the fitted model we find that the relationship between mean and variance is inconsistent with the distributional assumption. If the variance is larger than it should be from the implied relationship, then we say that the data exhibit *overdispersion*, while if it is smaller than it should be then the data exhibit *underdispersion*. The former is the more common in practice, typically arising as a result of some sort of clustering or clumping process. For example, in a traffic census, the number of cars passing the census point in a given time may well have a Poisson distribution, but if we actually count the number of *people* passing this point, then the presence of more than one person in some of the cars will lead to a set of data in which there is overdispersion relative to standard Poisson assumptions.

Failure to take account of overdispersion can lead to serious underestimation of standard errors and hence to misleading inferences about the regression parameters of the linear predictor. Various methods could be adopted for dealing with the problem, but one straightforward approach is to assume some more general form for the variance function than that given by standard theory, by incorporating an extra parameter. Quasi-likelihood can then be used to estimate the regression parameters, and some *ad hoc* estimation procedure for the extra overdispersion parameter. One such procedure, used already in other contexts, is to equate the Pearson X^2 statistic to its degrees of freedom and solve the resulting equation for the unknown parameter.

For example, in the case of overdispersed binomial data we can take

$$\text{Var}(Y) = \phi\mu(1 - \mu/n),$$

while in the case of overdispersed Poisson data we can take

$$\text{Var}(Y) = \phi\mu$$

where $\phi > 1$ in both cases. Using the quasi-likelihood method, we see that ϕ cancels from the esimating equations, so that the regression parameters remain the same as if ϕ were equal to 1. However, it can be shown that $\text{cov}(\hat{\beta})$ is proportional to ϕ, so standard errors are those of the ordinary model multiplied by the square root of the estimate of ϕ.

Taking the reasoning of this section one stage further, it may be the case that the dispersion parameter (or, more generally, the variance of the response variable) varies in a systematic way with the explanatory variables, independently of the variation of the mean with these variables. In this case we might want to model *both* mean and variance as functions of the explanatory variables. Essentially, this is done by specifying a second linear predictor, which is then applied to the dispersion parameter via a second link function and both components are joined together in a single model. However, the resulting complexities again take us well beyond the scope of the present book; more details are provided in Chapter 10 of *Generalized Linear Models* by P. McCullagh and J.A. Nelder (1989, Chapman & Hall).

8.5 Non-parametric models

All the models considered in this book have been concerned with either predicting the values of, or explaining the behaviour of, a response variable Y in terms of a linear combination $\eta = \beta_1 X_1 + \cdots + \beta_p X_p$ of explanatory variables $X_1, \ldots, X_p$. When the response variable can be assumed to be normal we have either the regression models of Chapter 3 or the analysis of variance models of Chapter 4, and in both these cases we model the mean of Y by the linear combination η. For other forms of response variable we need to use the generalized linear models of Chapters 5 to 7, and here we may need a link function to connect the mean of Y to the linear combination η. However, irrespective of which case we deal with, these are all *parametric* models depending on the parameters $\beta_1, \ldots, \beta_p$ of the linear combination. The procedure followed in all cases is to use some optimal method (such as maximum likelihood or least squares) for estimating these parameters given a set of data, and then to use the estimated parameters in either prediction or explanation.

With the advent of powerful computing capabilities in the 1970s and 1980s attention turned away from parametric to *data based* model building, and there has been much interest in non-parametric models in recent years. Such models abandon specified parametric functions of the explanatory variables, and simply look for the best 'smooth functions' of these variables that will provide good prediction or explanation of Y for the data in hand.

Instead of multiple regression, we might look for the best smooth function of the X_i to put through the mean of the observed y_i. An intuitively obvious way to do this is just to take the mean of Y replicates at each distinct combination of X_i values and join these means up in a smooth curve. Of course with small samples and/or large numbers of X variables there will only be one Y value at each distinct combination of the Xs, so the next best thing to do is to average y_i that are 'close' to each other in terms of their associated Xs. One general strategy for doing this is to define a *kernel* function which assigns weights to all sample observations based on their distances from each target individual, and then to take a weighted average of all observations at each sample point. This is the basis of 'locally smoothed' non-parametric regression. A second general strategy is to estimate the optimal function by separate functions in different portions of the data space, i.e. a *piecewise parametric* fitting procedure. The most flexible approach is to use different low-order polynomials (*splines*) in the subregions of space that are constructed as combinations of intervals (*knots*), with the imposition of a *roughness penalty* that trades off the smoothness of the fitted function against its faithfulness to the data. Both approaches have many technical features, problems and considerations which are detailed, for example, by P.J. Green and B.W. Silverman in the monograph *Non-parametric Regression and Generalized Linear Models: A Roughness Penalty Approach* (1983, Chapman & Hall).

Use of these ideas in extending generalized linear models is then fairly straightforward in principle. Instead of the linear predictor $\eta = \beta_1 X_1 + \cdots + \beta_p X_p$ we postulate the additive function $\eta = s_1(X_1) + \cdots + s_p(X_p)$, where $s_i(X_i)$ is an (unspecified) smooth function of X_i for $i = 1, \ldots, p$ and estimation of these functions can be conducted by means of local scoring using scatterplot smoothers. Once again, however, there are many technical aspects which cannot be covered here; the reader is referred to *Generalized Additive Models* by T.J. Hastie and R.J. Tibshirani (1990, Chapman & Hall) for a comprehensive account.

8.6   Conclusion: the art of model building

The preceding chapters have set out what is hopefully a fairly complete account of the range of models that are most useful in practical data analysis and, therefore, most likely to be included in degree courses whose major components include statistics. Almost inevitably therefore, the book will be used as a catalogue of the different models and their associated technicalities of estimation, goodness of fit and use in prediction or explanation. However, even though space restrictions have precluded a wider, more discursive, account of general aspects of model building for data analysis, it is hoped that some of the philosophy inherent in model building has nevertheless managed to seep out through the various illustrative examples that run through the book. In conclusion, therefore, it seems appropriate to provide some comments and reflections on this general process of model building.

The first major point to be made is that model building is an *iterative process* in which, typically, more than one person will be involved. Usually, the statistician will have been consulted by a researcher or scientist who has a specific problem to solve and who has collected some data to shed light on this problem. The statistician needs to extract the relevant features of the problem, provide a suitable model to fit to the data, and use the fitted model to solve the problem and answer the scientist's questions. An appropriate model may not be readily apparent, so the statistician may have to make a number of simplifying assumptions and perhaps start with a fairly tentative model. This can be fitted to the data and the fit critically examined using available diagnostics. If it seems satisfactory then the analysis could be carried out and the results discussed with the researcher. At this stage more information might come to light, or the researcher might be unhappy with some of the assumptions made, and the statistician would have to go back and rethink the model. The above cycle would then be repeated on a new model, as it would if any of the diagnostics showed that the current model did not fit the data at any stage. This process could continue for as many as half a dozen iterations before all parties were satisfied with the outcome.

Several allied points follow on from this one. The first is that model building is as much an art as a science. Although the actual *fitting* and *assessment* of a given model involves much sophisticated mathematics and computation, the process of extracting essential features of a problem and choosing the model to fit to the data is often a matter of fine judgement and subjectivity. *All models contain assumptions*, and it is important to match these assumptions to the physical characteristics of the problem as closely as possible. Sometimes a number of very different models all seem equally plausible in a given situation, and the modeller's experience may be critical in deciding between them.

The second allied point is that the modeller should always examine any fitted model very critically, to be sure that it provides a satisfactory basis for the analysis. There is little point in providing the scientist with what seem to be definite answers to questions if the foundations on which these answers are built turn out to be shaky.

Finally, perhaps the most important point to bear in mind is that the whole purpose of modelling is to provide *useful* answers to *practical* questions, and not just to conduct elegant mathematics. Fitting a sophisticated model may be an impressive technical accomplishment, but it is virtually worthless if it does not answer the questions posed or, worse still, if it cannot be understood by the scientist for whom it has been constructed. In the end, the true measure of success of any model is the extent to which it has met its practical objectives.

References

Agresti, A. (1996) *An Introduction to Categorical Data Analysis*, John Wiley & Sons, New York.

Aitkin, M., Anderson, D., Francis, B. and Hinde, J. (1989) *Statistical Modelling in GLIM*, Clarendon Press, Oxford.

Ashford, J.R. and Sowden, R.R. (1970) Multivariate probit analysis. *Biometrics*, **26**, 535-546.

Barnett, V. (1974) *Elements of Sampling Theory*, English Universities Press, London.

Bishop, Y.M.M. (1969) Full contingency tables, logits, and split contingency tables. *Biometrics*, **25**, 119-128.

Collett, D. (1991) *Modelling Binary Data*, Chapman & Hall, London.

Collett, D. (1994) *Modelling Survival Data in Medical Research*, Chapman & Hall, London.

Crowder, M.J., Kimber, A.C., Smith, R.L. and Sweeting, T.J. (1991) *Statistical Analysis of Reliability Data*, Chapman & Hall, London.

Draper, N.R. and Smith, H. (1981) *Applied Regression Analysis* (2nd Edition), John Wiley & Sons, New York.

Efron, B. (1986) Double exponential families and their use in generalized linear regression. *Journal of the American Statistical Association*, **81**, 709-721.

Everitt, B. (1994) *A Handbook of Statistical Analyses using S-plus*, Chapman & Hall, London.

Fienberg, S.E. (1980) *The Analysis of Cross-classified Categorical Data* (2nd Edition), The MIT Press, Cambridge, Massachusetts.

Finney, D.J. (1947) The estimation from original records of the relationship between dose and quantal response. *Biometrika*, **34**, 320-334.

Green, P.J. and Silverman, B.W. (1993) *Non-parametric Regression and Generalized Linear Models: A Roughness Penalty Approach*, Chapman & Hall, London.

Hald, A. (1952) *Statistical Theory with Engineering Applications*, John Wiley & Sons, New York.

Hastie, T.J. and Tibshirani, R.J. (1990) *Generalized Additive Models*, Chapman & Hall, London.

Healy, M.J.R. (1989) *GLIM: An Introduction*, Clarendon Press, Oxford.

Hewlett, P.S. and Plackett, R.L. (1950) Statistical aspects of the independent joint action of poisons, particularly insecticides. II. Examination of data for agreement with hypothesis. *Annals of Applied Biology*, **37**, 527-552.

Hinkley, D.V., Reid, N. and Snell, E.J. (Eds) (1991) *Statistical Theory and Modelling*, Chapman & Hall, London.

McCullagh, P. and Nelder, J.A. (1989) *Generalized Linear Models* (2nd Edition), Chapman & Hall, London.

Mead, R. and Curnow, R. N. (1983) *Statistical Methods in Agriculture and Experimental Biology*, Chapman & Hall, London.

Montgomery, D.C. and Peck, E.A. (1982) *Introduction to Linear Regression Analysis*, John Wiley & Sons, New York.

Muller, T.P. and Mayhall, J.T. (1971) Analysis of contingency table data on *torus mandibularis* using a loglinear model. *American Journal of Physical Anthropology*, **34**, 149-154.

Murdoch, J. and Barnes, J.A. (1970) *Statistical Tables for Science, Engineering, Management and Business Studies*, Macmillan, London.

Steel, R.G.D. and Torrie, J.H. (1980) *Principles and Procedures of Statistics, a Biometrical Approach* (2nd Edition), McGraw-Hill, New York.

Venables, W.N. and Ripley, B.D. (1995) *Modern Applied Statistics using S-plus*, Springer, New York.

Yule, G.U. (1900) On the association of attributes in statistics: with illustration from the material of the childhood society &c. *Philosophical Transactions of the Royal Society, Series A*, **194**, 257-319.

Index

.

Printed and bound by CPI Group (UK) Ltd, Croydon, CR0 4YY

27/10/2024

14580203-0001